Mathematics and Computing
Current Research and Developments

Mathematics and Computing
Current Research and Developments

Editors

Sk. Md. Abu Nayeem, J. Mukhopadhyay, S. B. Rao

Narosa Publishing House

New Delhi   Chennai   Mumbai   Kolkata

Mathematics and Computing
Current Research and Developments
160 pgs. | 34 figs. | 9 tbls.

Editors

Sk. Md. Abu Nayeem
J. Mukhopadhyay
Department of Mathematics
Aliah University, DN-41
Sector V, Salt Lake
Kolkata

S. B. Rao

(DST-CMS Project)
C R Rao Advanced Institute of Mathematics
Statistics & Computer Science
University of Hyderabad Campus
Hyderabad

NAROSA PUBLISHING HOUSE PVT. LTD.

22 Delhi Medical Association Road, Daryaganj, New Delhi 110 002
35-36 Greams Road, Thousand Lights, Chennai 600 006
306 Shiv Centre, Sector 17, Vashi, Navi Mumbai 400 703
2F-2G Shivam Chambers, 53 Syed Amir Ali Avenue, Kolkata 700 019

www.narosa.com

Printed from the camera-ready copy provided by the Editors.

ISBN 978-81-8487-321-4

Published by N.K. Mehra for Narosa Publishing House Pvt. Ltd.,
22 Delhi Medical Association Road, Daryaganj, New Delhi 110 002

Printed in India

Preface

The aim of this volume is to focus on theoretical and computational aspects of mathematics and its applications in different fields. Research papers have been selected from all areas of current research topics in mathematics including algebra, analysis, mathematical biology, fuzzy sets and their applications, fluid mechanics and astrophysics. We hope that this collection of selected papers presented in the National Seminar on "Current Research and Developments in Mathematics and Computing" which was held at Aliah University, Kolkata will be beneficial to the researchers, as well as the post graduate students of mathematics.

We are grateful to Prof. Syed Samsul Alam, Honourable Vice Chancellor of Aliah University for his constant support, encouragement and help in every respect in organizing the seminar and in bringing out this volume. We also gratefully acknowledge the financial support extended by DST, CSIR and DRDO for the event. Special thanks are due to the authority of Aliah University for sanctioning fund in organizing the event.

Lastly, we thank the officials, faculty members, staff and students of Aliah University for their effort to make the event successful.

Sk. Md. Abu Nayeem   J. Mukhopadhyay   S. B. Rao

Contents

Recent Developments of Sequences of Fuzzy Numbers in the Direction of Statistical Convergence

P.D. Srivastava* and S. Mohanta**

Department of Mathematics, Indian Institute of Technology, Kharagpur - 721 302, India.

Abstract. In this paper, we introduce and study the concept of Δ^m-summable sequence of fuzzy numbers by using a modulus function and the concept of Δ^m-statistical convergence of sequences fuzzy numbers. Also we have defined Δ^m-statistical pre-Cauchy sequences of fuzzy numbers.

Key words: Fuzzy number, modulus function, Δ^m-statistical convergence, Δ^m-statistical pre-Cauchy sequence.

1 Introduction

The objective "fuzzy" seems to be a very popular and very frequent one in the corresponding studies concerning the logical and set theoretical foundations of mathematics. The main reason for this quick development is, in our opinion, easy to be understood. The surrounding us, world is full of uncertainty, the information we obtain from the environment, the notions we use and the data resulting from our observation or measurement are, in general, vague and incorrect. So every formal description of the real world or some of its aspects is, in every case, only a approximation and the idealization of the actual state. The notions like fuzzy sets, fuzzy orderings, fuzzy languages etc. enable to handle and to study the degree of uncertainty mentioned above in a purely mathematical and formal way.

The concept of fuzzy sets and fuzzy set operations were first introduced by Zadeh [1] and subsequently several authors have discussed various aspects of the theory and applications of fuzzy sets such as topological spaces, similarity relations and fuzzy orderings, fuzzy mathematical programming. Many mathematicians have also applied the notion of fuzzy numbers on sequence spaces of scalars and studied their different properties such as Paranormed sequence spaces, difference sequence spaces, sequence spaces by using modulus functions, statistical convergence of sequences of fuzzy numbers etc. Matloka [2], Nanda [3], Nuray and Savas [4], Bilgin [5], Colak [6], Kwon [7] and several other authors studied the sequence spaces by using fuzzy numbers in an

---
* Corresponding author. Email: pds@maths.iitkgp.ernet.in, Telephone No.: 03222-283678(O), Fax No.: 03222-281718

** Email: sushomita@gmail.com

analogous way as Simons [8], Maddox [9], Kizmaz [10], Fast [11], Schoenberg [12], Salat [13], Fridy [14] and several authors studied for scalar valued sequence spaces.

The notion of statistical convergence is introduced by Fast [11] and Schoenberg [12] independently. Fast [11] introduced the idea of statistical convergence of Real or Complex numbers and Schoenberg [12] studied statistical convergence as a summability method and listed some of the properties of statistical convergence. From the point of view of sequence spaces, this concept has been generalized and developed by Salat [13], Fridy [14] and many others.

The idea of statistical convergence depends on the density of subsets of the set N of natural numbers.

The natural density of a subset K of N is defined by $\delta(K) = \lim_{n \to \infty} \frac{1}{n} |\{k \leq n : k \in K\}|$, where $|\{k \leq n : k \in K\}|$ denotes the number of elements of K not exceeding n.

If $X = (X_k)$ is a sequence that satisfies a property P for almost all k except a set of natural density zero, then we say that X_k satisfies P for almost all k and we write it by a.a.k.

The existing literature on statistical convergence appears to have been restricted to Real or Complex numbers, but Nuray and Savas [4] extended the idea to apply to sequences of fuzzy numbers. They introduced the concept of statistically convergent and statistically Cauchy sequences of fuzzy numbers and proved the equivalence relation between these two concepts. Later on, many authors such as Bilgin [5], Colak [6], Kwon [7] introduced different types of sequences of fuzzy numbers by using statistical convergence.

In continuation with this, we have introduced and studied the concept of strongly Δ^m-summable sequence of fuzzy numbers by using a modulus function and the concept of Δ^m-statistical convergence of sequences of fuzzy numbers where Δ^m-statistical pre-Cauchy sequences of fuzzy numbers is also discussed.

2   Definitions and Preliminaries

Fuzzy sets are considered with respect to a non-empty base space U of elements of interest. The essential idea is that each element $u \in U$, a membership grade $X(u)$ is assigned taking values in [0,1], with $X(u) = 0$ corresponding to nonmembership, $0 < X(u) < 1$ to partial membership and $X(u) = 1$ to full membership. According to Zadeh [1], a fuzzy subset of U is a non-empty subset $\{(u, X(u)) : u \in U\}$ of $U \times [0, 1]$ for some function $X : U \to [0, 1]$. The function X itself is often used for the fuzzy set.

Let $C(R^n) = \left\{ A \subset R^n : A \text{ is compact and convex set} \right\}$.

The space $C(R^n)$ has a linear structure induced by the operations

$$A + B = \left\{ a+b : a \in A, b \in B \right\} \text{ and } \lambda A = \left\{ \lambda a : a \in A \right\}$$

for $A, B \in C(R^n)$ and $\lambda \in R$.

The Hausdorff distance between A and B in $C(R^n)$ is defined by

$$\delta_\infty(A, B) = \max \left\{ \sup_{a \in A} \inf_{b \in B} ||a - b||, \sup_{b \in B} \inf_{a \in A} ||a - b|| \right\}.$$

It is well known $(C(R^n), \delta_\infty)$ is a complete metric space.

If R^n is replaced by R, then obviously the set $C(R^n)$ is reduced to the set of all closed bounded intervals on R, and so

$$d(A, B) = \max \left\{ |\underline{A} - \underline{B}|, |\overline{A} - \overline{B}| \right\}$$

for $A = [\underline{A}, \overline{A}]$, $B = [\underline{B}, \overline{B}] \in R$.

Definition 1 *(Diamond and Kloeden [15]) Denote $L(R) = \{X : R \to [0,1]$ such that*
X satisfies $(i) - (iv)$ below}, where

(i) X is normal, i.e., there exists a $t_0 \in R$ such that $X(t_0) = 1$.
(ii) X is fuzzy convex, i.e., $X(\lambda s + (1 - \lambda)t) \geq \min\{X(s), X(t)\}$ for all $s, t \in R$ and for all $\lambda \in [0,1]$.
(iii) X is upper semi-continuous.
(iv) The set $[X]^0 = \overline{\{t \in R : X(t) > 0\}}$ is compact, where $\overline{\{t \in R : X(t) > 0\}}$ denotes the closure of the set $\{t \in R : X(t) > 0\}$ in the usual topology of R.

The set $L(R)$ is called the set of all fuzzy numbers. The properties of fuzzy numbers imply that for each $\alpha \in [0,1]$, the α-level set X^α is a non-empty compact, convex subset of R with support $[X]^0$.

Definition 2 *The set $L(R)$ forms a linear space under addition and scalar multiplication in terms of α-level sets as defined below:*

$$[X + Y]^\alpha = [X]^\alpha + [Y]^\alpha \text{ and } [\lambda X]^\alpha = \lambda [X]^\alpha \quad \text{for each } 0 \leq \alpha \leq 1.$$

where X^α is given as

$$X^\alpha = \begin{cases} t : X(t) \geq \alpha & \text{if } \alpha \in (0,1] \\ t : X(t) > 0 & \text{if } \alpha = 0. \end{cases}$$

4   P.D. Srivastava and S. Mohanta

For each $\alpha \in [0,1]$, the set X^α is a closed, bounded and nonempty interval of R.

It is easy to verify that $\overline{d} : L(R) \times L(R) \to R$ be defined by

$$\overline{d}(X,Y) = \sup_{0 \le \alpha \le 1} d(X^\alpha, Y^\alpha).$$

is a metric on $L(R)$.

Definition 3 *(Matloka [2]) A sequence $X = (X_k)$ of fuzzy numbers is a function X from the set N of natural numbers into $L(R)$. The fuzzy number X_k denotes the value of the function at $k \in N$ and is called the k-th term of the sequence. Let $W(F)$ denote the set of all sequences of fuzzy numbers. The space $W(F)$ is a linear space of fuzzy numbers under coordinate wise addition and scalar multiplication.*

Definition 4 *(Colak et al. [6]) Let $X = (X_k)$ be a sequence of fuzzy numbers. Then the sequence $X = (X_k)$ is said to be Δ^m-bounded if the set $\{\Delta^m X_k : k \in N\}$ of fuzzy numbers is bounded, i.e., $\sup_k \overline{d}(\Delta^m X_k, \overline{0}) < \infty$.*

Definition 5 *(Colak et al. [6]) Let $X = (X_k)$ be a sequence of fuzzy numbers. Then the sequence $X = (X_k)$ is said to be Δ^m-convergent to the fuzzy number X_0, written as $\lim_{k \to \infty} \Delta^m X_k = X_0$, if for every $\varepsilon > 0$ there exists a positive integer k_0 such that $\overline{d}(\Delta^m X_k, X_0) < \varepsilon$ for all $k > k_0$.*

Definition 6 *A metric $\overline{d}$ on $L(R)$ is said to be translation invariant if $\overline{d}(X + Z, Y + Z) = \overline{d}(X,Y)$ for all $X, Y, Z \in L(R)$.*

Lemma 1 *(Mursaleen and Basarir [16]) If $\overline{d}$ is a translation invariant metric on $L(R)$, then*

(i) $\overline{d}(X + Y, \overline{0}) \le \overline{d}(X, \overline{0}) + \overline{d}(Y, \overline{0})$
(ii) $\overline{d}(\lambda X, \overline{0}) \le |\lambda| \, \overline{d}(X, \overline{0}), \quad |\lambda| > 1$.

3   New Sequence Space $W^F(\Delta^m, f, p)$

Let f be a modulus function and $p = (p_k)$ is a bounded sequence of strictly positive real numbers. Then we define the following sequence space $W^F(\Delta^m, f, p) =$
$$\left\{ X = (X_k) \in W(F) : \frac{1}{n} \sum_{k=1}^n \left(f(\overline{d}(\Delta^m X_k, L)) \right)^{p_k} \to 0 \text{ as } n \to \infty \text{ for some } L \right\}$$
where

$$\Delta^m X_k = \sum_{i=0}^m (-1)^i \binom{m}{i} X_{k+i}.$$

Theorem 1 $W^F(\Delta^m, f, p)$ *is a linear space over R where (p_k) be a bounded sequence of positive real numbers.*

Theorem 2 *Let (p_k) be a bounded sequence of positive real numbers such that $\inf p_k > 0$. Then the sequence space $W^F(\Delta^m, f, p)$ is a complete metric space with respect to the metric*

$$g(X,Y) = \sum_{i=1}^{m} f(\overline{d}(X_i, Y_i)) + \sup_n \left(\frac{1}{n} \sum_{k=1}^{n} \left(f(\overline{d}(\Delta^m X_k, \Delta^m Y_k)) \right)^{p_k} \right)^{\frac{1}{M}}$$

where $M = \max(1, \sup_k p_k)$.

Theorem 3 *Let $p = (p_k)$, $t = (t_k)$ be two sequences of positive real numbers and assume that for each $k \in N$, $0 < p_k \le t_k$ and the sequence $(\frac{t_k}{p_k})$ be bounded. Then $W^F(\Delta^m, f, t) \subset W^F(\Delta^m, f, p)$.*

Theorem 4 *Let f and g be two modulus functions. Then we have*

(i) $W^F(\Delta^m, f, p) \cap W^F(\Delta^m, g, p) \subseteq W^F(\Delta^m, f + g, p)$.
(ii) $W^F(\Delta^m, f, p) = W^F(\Delta^m, g, p)$ *if* $0 < \inf \frac{f(x)}{g(x)} \le \sup \frac{f(x)}{g(x)} < \infty$.

4 Δ^m-Statistical Convergence

Definition 7 *The sequence $X = (X_k)$ of fuzzy numbers is said to be Δ^m-statistically convergent to the fuzzy number L where $L \in L(R)$, if for every $\varepsilon > 0$,*

$$\lim_{n \to \infty} \frac{1}{n} |\{k \le n : \overline{d}(\Delta^m X_k, L) \ge \varepsilon\}| = 0.$$

Let $S^F(\Delta^m)$ denotes the set of all Δ^m-statistically convergent sequences of fuzzy numbers.

Definition 8 *The sequence $X = (X_k)$ of fuzzy numbers is said to be Δ^m-statistically Cauchy sequence, if for any $\varepsilon > 0$, there exists a positive integer k_0 (depends upon ε only) such that*

$$\lim_{n \to \infty} \frac{1}{n} |\{k \le n : \overline{d}(\Delta^m X_k, \Delta^m X_{k_0}) \ge \varepsilon\}| = 0.$$

Definition 9 *The sequence $X = (X_k)$ of fuzzy numbers is said to be Δ^m-statistically pre-Cauchy sequence, if for all $\varepsilon > 0$,*

$$\lim_{n \to \infty} \frac{1}{n^2} |\{(i, j) : i, j \le n, \overline{d}(\Delta^m X_i, \Delta^m X_j) \ge \varepsilon\}| = 0.$$

Remark 1 *If a sequence is Δ^m-convergent, then it is Δ^m-statistically convergent. But the converse is not true. This can be viewed from the following example.*

Example 1 *Let $m = 1$ and consider the sequence X as given below.*
When $k = 10^n$,

$$X_k(t) = \begin{cases} \frac{k}{k-1}(t+2-\frac{1}{k}) & \text{if } \frac{1-2k}{k} \le t \le -1 \\ \frac{k}{k+1}(\frac{1}{k}-t) & \text{if } -1 \le t \le \frac{1}{k} \\ 0 & \text{otherwise,} \end{cases}$$

and when $k \ne 10^n$,

$$X_k(t) = \begin{cases} t-5 & \text{if } 5 \le t \le 6 \\ 7-t & \text{if } 6 \le t \le 7 \\ 0 & \text{otherwise.} \end{cases}$$

The figure for the sequence (X_k) looks like as below:

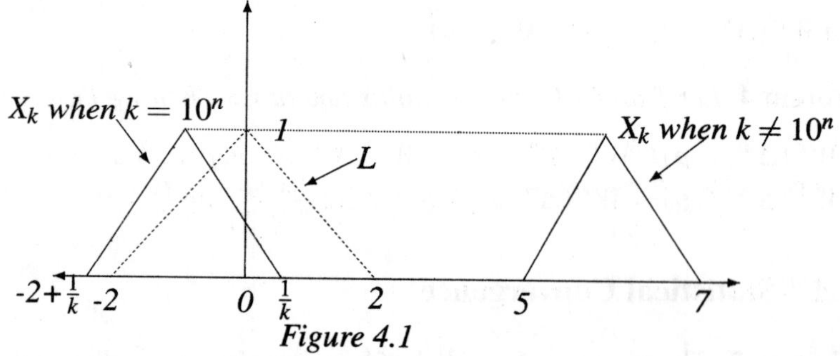

Figure 4.1

Then

$$[X_k]^\alpha = \begin{cases} [\frac{1-2k+k\alpha-\alpha}{k}, \frac{1-k\alpha-\alpha}{k}] & \text{when } k = 10^n \\ [5+\alpha, 7-\alpha] & \text{otherwise,} \end{cases}$$

i.e.,

$$[\Delta X_k]^\alpha = \begin{cases} [\frac{1-9k+2k\alpha-\alpha}{k}, \frac{1-2k\alpha-5k-\alpha}{k}] & \text{when } k = 10^n \\ [\frac{5k+2k\alpha+4+3\alpha}{k+1}, \frac{9k-2k\alpha+8-\alpha}{k+1}] & \text{when } k+1 = 10^n \\ [-2+2\alpha, 2-2\alpha] & \text{otherwise,} \end{cases}$$

which implies that $\Delta X_k \to L$ statistically, where $L = [-2+2\alpha, 2-2\alpha]$, but (ΔX_k) is not a convergent sequence.

Theorem 5 *Let f be any modulus function and $0 < h = \inf p_k \le p_k \le \sup p_k = H < \infty$. Then $W^F(\Delta^m, f, p) \subsetneq S^F(\Delta^m)$.*

Proof. Let $X \in W^F(\Delta^m, f, p)$ and $\varepsilon > 0$ be given. Then

$$\frac{1}{n}\sum_{k=1}^{n}\left(f(\overline{d}(\Delta^m X_k, L))\right)^{p_k} \ge \frac{1}{n}\sum_{\substack{k=1, \\ \overline{d}(\Delta^m X_k, L) \ge \varepsilon}}^{n}\left(f(\overline{d}(\Delta^m X_k, L))\right)^{p_k}$$

$$\ge \min(f(\varepsilon)^h, f(\varepsilon)^H)\frac{1}{n}|\{k \le n : \overline{d}(\Delta^m X_k, L) \ge \varepsilon\}|$$

which implies X is Δ^m-statistically convergent sequence.

Remark 2 *The inclusion is strict. This can be viewed from the following example.*

Example 2 *Let $f(x) = x$, $m = 1$, $p_k = 1$ for each $k \in N$ and consider the sequence X_k as given below.*
When $k = 5^n$,

$$X_k(t) = \begin{cases} k(t + \frac{1}{k}) & \text{if } \frac{-1}{k} \leq t \leq 0 \\ k(\frac{1}{k} - t) & \text{if } 0 \leq t \leq \frac{1}{k} \\ 0 & \text{otherwise,} \end{cases}$$

and when $k \neq 5^n$,

$$X_k(t) = \begin{cases} t - 5 & \text{if } 5 \leq t \leq 6 \\ 7 - t & \text{if } 6 \leq t \leq 7 \\ 0 & \text{otherwise.} \end{cases}$$

The figure for the sequence (X_k) looks like as below:

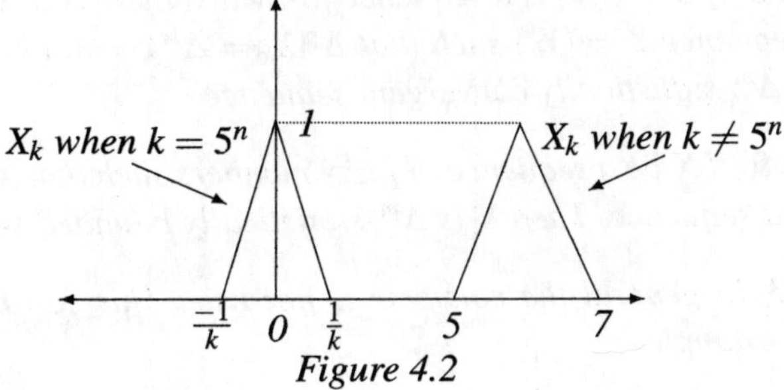

Figure 4.2

Then

$$[X_k]^{\alpha} = \begin{cases} [\frac{\alpha-1}{k}, \frac{1-\alpha}{k}] & \text{when } k = 5^n \\ [5 + \alpha, 7 - \alpha] & \text{otherwise,} \end{cases}$$

i.e.,

$$[\Delta X_k]^{\alpha} = \begin{cases} [\frac{\alpha-1-7k+\alpha k}{k}, \frac{1-5k-\alpha-\alpha k}{k}] & \text{when } k = 5^n \\ [\frac{5k+k\alpha+2\alpha+4}{k+1}, \frac{7k-k\alpha+8-2\alpha}{k+1}] & \text{when } k + 1 = 5^n \\ [-2 + 2\alpha, 2 - 2\alpha] & \text{otherwise.} \end{cases}$$

Then $\Delta X_k \to L$ statistically, where $L = [-2 + 2\alpha, 2 - 2\alpha]$, but $(\Delta X_k) \notin w^F(\Delta^m, f, p)$.

Theorem 6 *If f is a bounded modulus function, then $S^F(\Delta^m) \subseteq W^F(\Delta^m, f, p)$.*

Proof. Let $\varepsilon > 0$ be given and f be any modulus function. Since f is a bounded modulus function, there exists an integer K such that $f(x) < K$ for all $x \geq 0$.

Let X is Δ^m-statistically convergent sequence. Consider

$$\frac{1}{n}\sum_{k=1}^{n}\left(f(\overline{d}(\Delta^m X_k, L))\right)^{p_k} = \frac{1}{n}\sum_{\substack{k=1,\\ \overline{d}(\Delta^m X_k, L)\geq\varepsilon}}^{n}\left(f(\overline{d}(\Delta^m X_k, L))\right)^{p_k}$$

$$+\frac{1}{n}\sum_{\substack{k=1,\\ \overline{d}(\Delta^m X_k, L)<\varepsilon}}^{n}\left(f(\overline{d}(\Delta^m X_k, L))\right)^{p_k}$$

$$\leq \max(K^h, K^H)\frac{1}{n}|\{k\leq n : \overline{d}(\Delta^m X_k, L)\geq\varepsilon\}|$$

$$+\max(f(\varepsilon)^h, f(\varepsilon)^H)$$

$$\to 0 \text{ as } n\to\infty.$$

i.e., $X \in W^F(\Delta^m, f, p)$ which implies $S^F(\Delta^m) \subseteq W^F(\Delta^m, f, p)$.

Theorem 7 *If the sequence X is Δ^m-statistically convergent, then X is Δ^m-statistically Cauchy.*

Theorem 8 *If $X = (X_k)$ is a sequence for which there is a Δ^m-statistically convergent sequence $Y = (Y_k)$ such that $\Delta^m X_k = \Delta^m Y_k$ a.a.k. Then the sequence X is also Δ^m-statistically convergent sequence.*

Theorem 9 *If X be a sequence of fuzzy numbers such that X is Δ^m-statistically convergent sequence. Then X is Δ^m-statistically bounded sequence.*

Remark 3 *In general the converse is not true. This can be viewed from the following example.*

Example 3 *Let $f(x) = x$, $m = 1$, $p_k = 1$ for each $k \in N$ and consider the sequence X_k as given below.*
When $k = 10^n$,

$$X_k(t) = \begin{cases} kt+1 & \text{if } \frac{-1}{k}\leq t\leq 0 \\ 1-kt & \text{if } 0\leq t\leq\frac{1}{k} \\ 0 & \text{otherwise,} \end{cases}$$

when $k \neq 10^n$ and k is odd,

$$X_k(t) = \begin{cases} t+7 & \text{if } -7\leq t\leq -6 \\ -t-5 & \text{if } -6\leq t\leq -5 \\ 0 & \text{otherwise,} \end{cases}$$

and when $k \neq 10^n$ and k is even

$$X_k(t) = \begin{cases} t-5 & \text{if } 5\leq t\leq 6 \\ 7-t & \text{if } 6\leq t\leq 7 \\ 0 & \text{otherwise.} \end{cases}$$

The figure for the sequence (X_k) looks like as below:

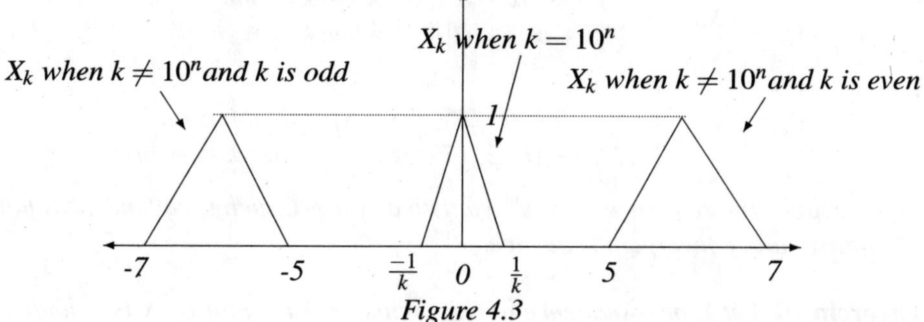

Figure 4.3

Then

$$[X_k]^\alpha = \begin{cases} [\frac{\alpha-1}{k}, \frac{1-\alpha}{k}] & \text{when } k = 10^n \\ [-7+\alpha, -5-\alpha] & \text{when } k \neq 10^n \text{ and } k \text{ is odd} \\ [5+\alpha, 7-\alpha] & \text{when } k \neq 10^n \text{ and } k \text{ is even} \end{cases}$$

i.e.,

$$[\Delta X_k]^\alpha = \begin{cases} [\frac{\alpha-1+\alpha k+5k}{k}, \frac{1-\alpha+7k-\alpha k}{k}] & \text{when } k = 10^n \\ [\frac{-7k+k\alpha+2\alpha-8}{k+1}, \frac{-5k-k\alpha-4-2\alpha}{k+1}] & \text{when } k+1 = 10^n \\ [-14+2\alpha, -10-2\alpha] & \text{when } k \neq 10^n \text{ and } k \text{ is odd} \\ [10+2\alpha, 14-2\alpha] & \text{when } k \neq 10^n \text{ and } k \text{ is even} \end{cases}$$

which implies X is Δ^m-statistically bounded sequence, but not Δ^m-statistically convergent sequence.

Remark 4 *A sequence X is Δ^m-statistically pre-Cauchy sequence, but not Δ^m-statistically convergent sequence.*

This can be viewed from the following example.

Example 4 *Let $f(x) = x$, $p_k = 1$ for each $k \in N$ and consider the sequence X_k as follows.*

When k is odd,

$$X_k(t) = \begin{cases} t+7 & \text{if } -7 \leq t \leq -6 \\ -t-5 & \text{if } -6 \leq t \leq -5 \\ 0 & \text{otherwise,} \end{cases}$$

and when k is even,

$$X_k(t) = \begin{cases} t-5 & \text{if } 5 \leq t \leq 6 \\ 7-t & \text{if } 6 \leq t \leq 7 \\ 0 & \text{otherwise.} \end{cases}$$

Then

$$[X_k]^{\alpha} = \begin{cases} [-7+\alpha, -\alpha-5] & \text{when } k \text{ is odd} \\ [5+\alpha, 7-\alpha] & \text{when } k \text{ is even,} \end{cases}$$

i.e.,

$$[\Delta^m X_k]^{\alpha} = \begin{cases} [2^m(-7+\alpha), 2^m(-\alpha-5)] & \text{when } k \text{ is odd} \\ [2^m(5+\alpha), 2^m(7-\alpha)] & \text{when } k \text{ is even,} \end{cases}$$

which implies the sequence X is Δ^m-statistically pre-Cauchy sequence, but not Δ^m-statistically convergent sequence.

Theorem 10 *Let X be a sequence of fuzzy numbers such that $\Delta^m X$ is bounded. Then X is said to be Δ^m-statistically pre-Cauchy if and only if*

$$\lim_{n\to\infty} \frac{1}{n^2} \sum_{i,j \leq n} f(\overline{d}(\Delta^m X_i, \Delta^m X_j)) = 0,$$

for any bounded modulus function f.

Proof. Let $\displaystyle\lim_{n\to\infty} \frac{1}{n^2} \sum_{i,j \leq n} f(\overline{d}(\Delta^m X_i, \Delta^m X_j)) = 0.$

Given $\varepsilon > 0$ and for any $n \in N$, we have

$$\frac{1}{n^2} \sum_{i,j \leq n} f(\overline{d}(\Delta^m X_i, \Delta^m X_j)) \geq \frac{1}{n^2} \sum_{\substack{i,j \leq n \\ \overline{d}(\Delta^m X_i, \Delta^m X_j) \geq \varepsilon}} f(\overline{d}(\Delta^m X_i, \Delta^m X_j))$$

$$\geq f(\varepsilon) \frac{1}{n^2} |\{(i,j) : i,j \leq n, \overline{d}(\Delta^m X_i, \Delta^m X_j) \geq \varepsilon\}|$$

and thus X is Δ^m-statistically pre-Cauchy sequence.

Conversely, let X is Δ^m-statistically pre-Cauchy sequence and $\varepsilon > 0$ be given. Choose $\delta > 0$ such that $f(\delta) < \frac{\varepsilon}{2}$. Since f is a bounded modulus function, so there exist an integer B such that $f(\overline{d}(\Delta^m X_i, \Delta^m X_j)) < B$. Now for each $n \in N$, consider

$$\frac{1}{n^2} \sum_{i,j \leq n} f(\overline{d}(\Delta^m X_i, \Delta^m X_j)) = \frac{1}{n^2} \sum_{\substack{i,j \leq n \\ \overline{d}(\Delta^m X_i, \Delta^m X_j) < \delta}} f(\overline{d}(\Delta^m X_i, \Delta^m X_j))$$

$$+ \frac{1}{n^2} \sum_{\substack{i,j \leq n \\ \overline{d}(\Delta^m X_i, \Delta^m X_j) \geq \delta}} f(\overline{d}(\Delta^m X_i, \Delta^m X_j))$$

$$\leq f(\delta) + B \frac{1}{n^2} |\{(i,j) : i,j \leq n, \overline{d}(\Delta^m X_i, \Delta^m X_j) \geq \delta\}|$$

$$\leq \frac{\varepsilon}{2} + B \frac{1}{n^2} |\{(i,j) : i,j \leq n, \overline{d}(\Delta^m X_i, \Delta^m X_j) \geq \delta\}|$$

Since let X is Δ^m-statistically pre-Cauchy sequence, so we have

$$\frac{1}{n^2}|\{(i,j) : i,j \le n, \overline{d}(\Delta^m X_i, \Delta^m X_j) \ge \delta\}| \to 0 \text{ as } n \to \infty.$$

i.e., there exist $n_0 \in N$ such that

$$\frac{1}{n^2}|\{(i,j) : i,j \le n, \overline{d}(\Delta^m X_i, \Delta^m X_j) \ge \delta\}| < \frac{\varepsilon}{2B} \text{ for all } n \ge n_0.$$

i.e.,

$$\frac{1}{n^2}\sum_{i,j\le n} f(\overline{d}(\Delta^m X_i, \Delta^m X_j)) \le \varepsilon \text{ for all } n \ge n_0.$$

Hence we have $\displaystyle\lim_{n\to\infty}\frac{1}{n^2}\sum_{i,j\le n} f(\overline{d}(\Delta^m X_i, \Delta^m X_j)) = 0.$

References

1. L.A. Zadeh, Fuzzy sets, Information Control, 8 (1965), 338–353.
2. M. Matloka, Sequences of fuzzy numbers, BUSEFAL, 28 (1986), 28–37.
3. S. Nanda, On sequences of fuzzy numbers, Fuzzy Sets and Systems, 33 (1989), 123–126.
4. F. Nuray and E. Savas, Statistical convergence of fuzzy numbers, Mathematica Slovaca, 45(3) (1995), 269–273.
5. T. Bilgin, Δ-statistical and strong Δ-Cesaro convergence of sequence of fuzzy numbers, Math. Commun., 8(1) (2003), 95–100.
6. R. Colak, Y. Altin and M. Mursaleen, On some sets of difference sequences of fuzzy numbers, Soft Comput., 15 (2011), 787–793.
7. J.S. Kwon, On statistical and p-Cesaro convergence of fuzzy numbers, Korean J. Comput. Appl. Math., 7(1) (2000), 195–203.
8. S. Simons, The sequence spaces $l(p_\mu)$ and $m(p_\mu)$, Proc. London Math. Soc., 15(3) (1965), 422–436.
9. I.J. Maddox, Spaces of strongly summable sequences, Quart. J. Math. Oxford, 2(18) (1967), 345–355.
10. H. Kizmaz, On certain sequence spaces, Can. Math. Bull., 24 (1981), 169–176.
11. H. Fast, Sur la convergence statique, Colloquium Math., 2 (1951), 241–244.
12. I.J. Schoenberg, The integrability of certain functions and related summability methods, Amer. Math. Monthly, 66 (1959), 361–375.
13. T. Salat, On Statistically convergent sequences of real numbers, Math. Slovaca, 30(2) (1980), 139–150.
14. J. Fridy, On Statistical convergence, Analysis, 5(4) (1985), 301–313.
15. P. Diamond and P. Kloeden, Metric Spaces of Fuzzy Sets, Elsevier, 1994.
16. M. Mursaleen and M. Basarir, On some new sequence spaces of fuzzy numbers, Indian J. Pure Appl. Math., 34(9) (2003), 1351–1357.

Dynamical Behaviour of a Single Species Population Model in a Polluted Environment

G.P. Samanta* and Swarnali Sharma**

Department of Mathematics, Bengal Engineering and Science University, Shibpur
Howrah - 711 103, India.

Abstract. In this paper we have discussed the dynamical behaviour of a single species population model in a polluted environment which describes the effect of toxicants on ecological system. Boundedness, positivity, stability analysis of the model at various equilibrium points are discussed thoroughly. We have also studied the effect of double discrete delays on the population model. The stability of the model with double time delays is investigated by the Nyquist criteria. By choosing one of the delays as a bifurcation parameter, the model is found to undergo a Hopf-bifurcation. Some numerical simulations for justifying the theoretical results are also illustrated by using MATLAB.

Key words: Boundedness, local and global Stability, Lyapunov function, double discrete delays, Nyquist criteria, Hopf-bifurcation.

1 Introduction

One of the greatest problem that the world is facing today is the change in the environment caused by pollution, affecting the long term survival of species, human life style and bio diversity of the habitat. Therefore, the study of the effects of toxicant on the population and the assessment of the risk to populations are becoming quite important in recent years. The problem of estimating qualitatively the effect of a toxicant on a population by mathematical models is a very effective way. This kind of population model was proposed by Hallam and his colleagues in 1980s [1–3]. This model was revised by many researchers [4–7]. Buonomo *et al.* [8] had studied the effect of variation of the population on the toxicant concentration in the organism and environment. There are many other works on the effects of a single toxicant on various ecosystem by using mathematical models [9–15]. Time delays have been incorporated in ecological models by many researchers [16–27]. Introducing such delays in ecological models make them more realistic and reasonable. These delay differential equations exhibit much more complicated dynamics than ordinary differential equations since a time delay could cause a stable equilibrium to become unstable and cause the population to fluctuate.

* Corresponding author. E-mail: g_p_samanta@yahoo.co.uk, gpsamanta@math.becs.ac.in
** E-mail: swarnali.sharma87@gmail.com

In this paper, we have developed a single species population model in polluted environment. In section 2, we present a brief sketch of the construction of the model. Boundedness and positivity analysis of the solutions are shown which implies that the system is ecologically well-behaved. In the next section, we have discussed the existence and stability analysis of the model at various equilibrium points under zero-exogeneous input and non-zero exogeneous input. Then we obtain necessary conditions for the existence of interior equilibrium E^* and local and global stability of the system at E^*. The analysis of double delayed model is described in section 4. The stability of the model with double delays is investigated by the Nyquist criteria. By choosing one of the delays as a bifurcation parameter, the model is found to undergo a Hopf-bifurcation. Some of the important analytic results are numerically verified by using MATLAB in the section 5. Finally, section 6 contains the general discussions of the paper and ecological implications of our mathematical findings.

2 Model Construction

In this paper we analyze a model which describes the effect of toxicant on a single species. Now, we shall discuss about the construction of our model [8]. Let Ω be a bounded domain in $\mathbb{R}^N (N = 1, 2, 3)$ with smooth boundary $\partial\Omega$. Let V be an arbitrary domain contained in Ω. The amount of population living in V at time t is denoted by $N_V(t)$, the total amount of a given (external) toxicant in V at time t is denoted by $S_V(t)$ and the total amount of the toxicant stored inside the bodies of the organisms living in V at time t is denoted by $Y_V(t)$. We assume that there exist three smooth functions $n(x,t), c(x,t), s(x,t)$ defined in $\Omega \times (0,T], (T \leq +\infty)$ such that for any V in Ω it results:

$$N_V(t) = \int n(x,t)dx; \quad S_V(t) = \int s(x,t)dx; \quad Y_V(t) = \int c(x,t)n(x,t)dx.$$

We assume that the population dynamics is governed by:

$$\dot{N}_V = \int_V [ng(n,c) - div J_n]dx,$$

where J_n is the flux of the population through the boundary of V. The growth rate $g(n,c)$ of the population is a smooth function given by the difference between the birth rate $b(n,c)$ and the death rate $d(n,c)$.
For the external toxicant we have,

$$\dot{S}_V = \int_V [-ksn + \{r + d(n,c)\}cn - hs + u - div J_s]dx.$$

The term $-ksn$ represents the uptake of the toxicant from the environment by population; rcn is the toxicant input to the environment from the population due to egestion; $d(n,c)cn$ represents the amount of toxicant stored by the living organisms which die at time t; hs denotes losses from the environment outside the system; u is the exogeneous toxicant input rate which is assumed to be a smooth bounded non-negative function of x and t.

For the internal toxicant we have,

$$\dot{Y}_V = \int_V [ksn - \{r+m+d(n,c)\}cn - divJ_y]dx.$$

The term $-mcn$ represents metabolization processes and other losses, J_z and J_y denote, respectively, the flux of external and internal toxicant through the boundary of V.

We assume a Fickian diffusion mechanism for the population and the external toxicant, that is,

$$J_n = -d_n\nabla n, \quad J_s = -d_s\nabla s,$$

where d_n and d_s are the diffusion coefficients (positive constants). As for the flux J_y, since the internal toxicant is in some sense drifted by the random walk of the living population, it seems rather natural to assume $J_y = cJ_n = -d_nc\nabla n$. Concerning the growth rate of the population we assume that the birth rate is $b(n,c) = b_0 - fn$ and the death rate is $d(n,c) = d_0 + \alpha c$, where b_0, d_0 and α are positive constants and f is assumed to be a non-negative constant. Therefore we assume, in absence of toxicant, a malthusian $(f = 0)$ or a logistic growth rate $(f > 0)$.

Because of the arbitrary of V and using standard regularity arguments, the following system of partial differential equation arises:

$$\frac{\partial n}{\partial t} - d_n\triangle n = n(b_0 - d_0 - \alpha c - fn)$$

$$\frac{\partial cn}{\partial t} - d_nc\triangle n - d_n\nabla c\nabla n = ksn - (r+m+d_0+\alpha c)cn \qquad (I)$$

$$\frac{\partial s}{\partial t} - d_s\triangle s = -ksn + (r+d_0+\alpha c)cn - hs + u$$

in $\Omega \times (0,T]$.

The second equation in (I) can be written as

$$n\frac{\partial c}{\partial t} + c\frac{\partial n}{\partial t} - d_nc\triangle n - d_n\nabla c\nabla n = ksn - (r+m+d_0+\alpha c)cn \qquad (II)$$

Thus, system (I) is equivalent to

$$\frac{\partial n}{\partial t} - d_n \triangle n = n(b_0 - d_0 - \alpha c - fn)$$

$$n\frac{\partial c}{\partial t} - d_n \nabla c \nabla n = ksn - (r + m + b_0 - fn)cn \qquad (III)$$

$$\frac{\partial s}{\partial t} - d_s \triangle s = -ksn + (r + d_0 + \alpha c)cn - hs + u$$

in $\Omega \times (0, T]$.

We further suppose no flux through the boundry $\partial \Omega$. Precisely, we append to system (I) homogeneous Neumann boundary conditions

$$\frac{\partial n}{\partial v} = \frac{\partial s}{\partial v} = 0, \quad \text{on } \partial \Omega \times (0, T].$$

Finally, we set initial data as:

$$n(x,0) = n_0(x) \geq 0; \quad c(x,0) = c_0(x) \geq 0; \quad s(x,0) = s_0(x) \geq 0; \quad x \in \bar{\Omega}.$$

If the initial data are constants then the system (III) reduces to the ordinary differential equation (ODE) system:

$$\frac{dn}{dt} = n(b_0 - d_0 - \alpha c - fn),$$

$$\frac{dc}{dt} = ks - (r + m + b_0 - fn)c, \qquad (IV)$$

$$\frac{ds}{dt} = -ksn + (r + d_0 + \alpha c)cn - hs + u.$$

2.1   Basic mathematical model

Here we assume:

   $n(t)$: concentration of the population biomass,

   $c(t)$: concentration of the toxicant in the population,

   $s(t)$: concentration of the toxicant in the environment.

The model satisfies the following assumptions:

(A1) There is a given toxicant in the environment and the living organisms absorb into their bodies part of this toxicant so that the dynamics of the population is affected by the toxicant.

(A2) For the growth rate of population we assume that the birth rate is $b_0 - fn(t)$ and the death rate is $d_0 + \alpha c(t)$, where b_0, f, d_0, α are assumed to be positive constants.

We consider the model:

$$\frac{dn}{dt} = n(b_0 - d_0 - \alpha c - fn),$$

$$\frac{dc}{dt} = ks - (r + m + b_0 - fn)c, \tag{1}$$

$$\frac{ds}{dt} = -ksn + (r + d_0 + \alpha c)cn - hs + u(t),$$

with initial data $n(0) \geq 0, c(0) \geq 0, s(0) \geq 0$.

Here

k : depletion rate of toxicant in the environment due to its intake made by the population,

r : depletion rate of toxicant in the population due to egestion,

m : depletion rate of toxicant in the population due to metabolization process,

h : depletion rate of toxicant in the environment,

$u(t)$: exogeneous toxicant input rate which is assumed to be a smooth bounded non-negative function of t.

We can see that, if $b_0 - d_0 - \alpha c(t) \leq 0$, then $n(t)$ will be going to extinct. So we suppose,

$$p = b_0 - d_0 > 0 \text{ and } b_0 - d_0 - \alpha c(t) > 0 \text{ implies } c(t) < \frac{b_0 - d_0}{\alpha}, \forall t \geq 0. \tag{2}$$

The model we have just specified has nine parameters, which make the analysis difficult. To reduce the number of parameters and to determine which combinations of parameters control the behaviour of the system, let us choose $X = fn, Y = \alpha c, Z = k\alpha s$.

So, system (1) becomes:

$$\frac{dX}{dt} = X(p - Y - X),$$

$$\frac{dY}{dt} = Z - (q - X)Y, \tag{3}$$

$$\frac{dZ}{dt} = -aXZ + a(d + Y)XY - hZ + \gamma(t),$$

with initial data:

$$X(0) \geq 0, Y(0) \geq 0, Z(0) \geq 0, \tag{4}$$

where

$p = b_0 - d_0, q = r + m + b_0, a = \frac{k}{f}, d = r + d_0, \gamma(t) = k\alpha u(t.$
Therefore from (2), we have

$$Y(t) < p, \quad \forall t \geq 0. \quad [\because Y(t) = \alpha c(t)] \tag{5}$$

Theorem 1. *Each component of the solution of system (3) subject to initial conditions are positive and bounded for all $t \geq \bar{t}$, where $\bar{t} = \inf\{t > 0 : X(t) > 0, Y(t) > 0, Z(t) > 0\}$.*

Proof. Here

$$\bar{t} = \inf\{t > 0 : X(t) > 0, Y(t) > 0, Z(t) > 0\}.$$

Thus $\bar{t} > 0$ and it follows from the first equation of the system (3) that

$$X(t) = X(\bar{t})\exp\left\{\int_{\bar{t}}^{t} p - Y(s) - X(s)ds\right\} > 0.$$

From the second equation of (3) we get,

$$\frac{dY}{dt} \geq -(q - X)Y$$

$$\Rightarrow Y(t) \geq Y(\bar{t})\exp\left\{-\int_{\bar{t}}^{t}(q - X(s))ds\right\} > 0.$$

Similarly, from the third equation of (3) we get,

$$\frac{dZ}{dt} \geq -(aX + h)Z$$

$$\Rightarrow Z(t) \geq Z(\bar{t})\exp\left\{-\int_{\bar{t}}^{t}(aX(s) + h)ds\right\} > 0.$$

Therefore, $X(t) > 0, Y(t) > 0, Z(t) > 0, \quad \forall t \geq \bar{t} > 0.$
From the first equation of (3), we have,

$$\frac{dX}{dt} = X(p - Y - X) \leq pX\left(1 - \frac{X}{p}\right).$$

By a standard comparison theorem and by (5), we have,

$$\limsup_{t \to \infty} X(t) \leq p, \quad Y(t) < p, \quad \forall t \geq 0.$$

From the third equation of (3), we have,

$$\frac{dZ}{dt} \leq a(d + p)p^2 - hZ + \gamma^M.$$

Therefore,

$$\limsup_{t\to\infty} Z(t) \leq \frac{a(d+p)p^2 + \gamma^M}{h}, \quad \text{where} \quad \gamma^M = \limsup_{t\to\infty} \gamma(t).$$

From the above discussion, we conclude that each component of the solution of system (3) subject to (4) is positive and bounded for all $t \geq \bar{t} > 0$.

This completes the proof.

3   Stability Behaviour of the Model

Case I: Zero exogeneous input $(\gamma(t) = 0)$

Theorem 2. *If $\gamma(t) = 0$, then the model (3) has non-negative equilibria $E_0(0, 0,0)$ which is unstable and $E_1(p,0,0)$ which is locally asymptotically stable. The interior equilibrium $E^*(X^*,Y^*,Z^*)$ is not feasible.*

Proof. The variational matrix of system (3) at E_0 is

$$V(E_0) = \begin{bmatrix} p & 0 & 0 \\ 0 & -q & 1 \\ 0 & 0 & -h \end{bmatrix}.$$

The eigenvalues are $p, -q, -h$. So, it is obvious that, E_0 is unstable (hyperbolic saddle).

Now the variational matrix of system (3) at E_1 is

$$V(E_1) = \begin{bmatrix} -p & -p & 0 \\ 0 & -(q-p) & 1 \\ 0 & adp & -(ap+h) \end{bmatrix}.$$

The characteristic equation of $V(E_1)$ is $(p+\lambda)(\lambda^2 + B\lambda + C) = 0$, where $B = ap+h+q-p = ap+h+d_0+r+m > 0$, and $C = (ap+h)(q-p)-apd = ap(q-p-d)+h(q-p) = apm+h(d_0+r+m) > 0$, since $p > 0$.

The eigenvalues are $\lambda_1 = -p < 0$ and $\lambda_{2,3} = \frac{-B\pm\sqrt{B^2-4C}}{2}$.

Since $B > 0$, $C > 0$, therefore the signs of the real parts of λ_2, λ_3 are negative. Hence E_1 is locally asymptotically stable.

It is noted here that the other equilibrium point (X^*,Y^*,Z^*) is not feasible, since

$$X^* = \frac{hq}{h-am}, \quad Y^* = \frac{-h(r+m+d_0)-apm}{h-am}$$

are opposite in sign since $p > 0$, by (2).

This completes the proof.

Case II: Non-zero exogeneous input $(\gamma(t) = Q > 0)$

When $(\gamma(t) = Q > 0)$, the model (3) has two non-negative equilibria, $E_2\left(0, \frac{Q}{hq}, \frac{Q}{h}\right)$ and $E^*(X^*, Y^*, Z^*)$. The variational matrix of system (3) at E_2 is given by

$$V(E_2) = \begin{bmatrix} p - \frac{Q}{hq} & 0 & 0 \\ \frac{Q}{hq} & -q & 1 \\ -\frac{aQ}{h} + \frac{aQ}{hq}\left(d + \frac{Q}{hq}\right) & 0 & -h \end{bmatrix}.$$

The characteristic equation of $V(E_2)$ is $\left(p - \frac{Q}{hq} - \lambda\right)(q + \lambda)(h + \lambda) = 0$.

So, E_2 is asymptotically stable if and only if $hpq < Q$ and if $hpq > Q$, then E_2 becomes unstable.

The interior equilibrium point $E^*(X^*, Y^*, Z^*)$ of system (3) is given by

$$X^* = \frac{A - hq + p(am - h)}{2(am - h)}, Y^* = \frac{p(am - h) + hq - A}{2(am - h)},$$

$$Z^* = \frac{\{p(am - h) + hq - A\}\{(am - h)(2q - p) + hq - A\}}{4(am - h)^2}$$

where

$$A = \sqrt{\{hq - p(am - h)\}^2 + 4(am - h)(hpq - Q)} \quad \text{and} \quad m = (q - p - d).$$

It can be provided that the unique interior equilibrium point $E^*(X^*, Y^*, Z^*)$ of system (3) exists if and only if the following two conditions are satisfied

$$(i)\ am > h \quad \text{and} \quad (ii)\ hpq > Q$$

Summarizing the above analysis we come to the following theorem:

Theorem 3. *If $\gamma(t) = Q > 0$, then the equilibrium point $E_2(0, \frac{Q}{hq}, \frac{Q}{h})$ of the system (3) is locally asymptotically stable if and only if $hpq < Q$ and the unique interior equilibrium point $E^*(X^*, Y^*, Z^*)$ of system (3) exists if and only if the following two conditions are satisfied*

$$(i)\ am > h \quad \text{and} \quad (ii)\ hpq > Q, \ m = (q - p - d).$$

3.1   Local stability of $E^*(X^*, Y^*, Z^*)$

The variational matrix at E^* is given by

$$V(E^*) = \begin{bmatrix} m_{11} & m_{12} & 0 \\ m_{21} & m_{22} & m_{23} \\ m_{31} & m_{32} & m_{33} \end{bmatrix},$$

where,

$$m_{11} = -X^*, \quad m_{12} = -X^*, \quad m_{21} = Y^*, \quad m_{22} = -q + X^*, \quad m_{23} = 1,$$

$$m_{31} = -aZ^* + aY^*(d + Y^*), \quad m_{32} = aX^*(d + 2Y^*), \quad m_{33} = -(aX^* + h).$$

The characteristic equation is

$$\lambda^3 + A_1\lambda^2 + A_2\lambda + A_3 = 0$$

where

$$A_1 = -m_{11} - m_{22} - m_{33} = -\mathrm{tr}\left[V(E^*)\right]$$

$$A_2 = m_{11}m_{22} + m_{11}m_{33} + m_{22}m_{33} - m_{23}m_{32} - m_{12}m_{21}$$

$$A_3 = -\det\left[V(E^*)\right] = m_{11}m_{23}m_{32} + m_{12}m_{21}m_{33} - m_{11}m_{22}m_{33} - m_{12}m_{23}m_{31}.$$

By the Routh-Hurwitz criterion [28–30], it follows that all eigenvalues of characteristic equation have negative real part if and only if

$$A_1 > 0, \quad A_3 > 0, \quad A_1A_2 - A_3 > 0. \tag{6}$$

Theorem 4. *E^* is locally asymptotically stable if and only if the inequalities (6) are satisfied.*

Further the conditions arising through Hopf-bifurcation can not be derived explicitly in terms of system parameters. Let us assume μ stands for any parameter involved with the model system (3). If there exists a critical magnitude $\mu = \mu^*$ such that $A_1(\mu^*) > 0, A_3(\mu^*) > 0, (A_1A_2 - A_3)_{\mu=\mu_*} = 0$ and $\frac{d}{d\mu}(A_1A_2 - A_3)_{\mu=\mu_*} \neq 0$, then E^* undergoes a Hopf-bifurcation at $\mu = \mu^*$ (Liu's criterion) [28–30].

3.2 Global stability of E^*

E^* is not always globally asymptotically stable. The conditions which guarantee the global stability of E^* are stated in the following theorem.

Theorem 5. *Since $X(t)$, $Y(t)$, $Z(t)$ are bounded, There exist positive constants m_i, M_i (i=1,2,3) such that $m_1 \leq X(t) \leq M_1$, $m_2 \leq Y(t) \leq M_2$, $m_3 \leq Z(t) \leq M_3$. If the following inequalities hold :*

$$(M_2 - 1)^2 < (q - X^*)$$

$$\{aM_2(d + M_2) - am_3\}^2 < (aX^* + h) \tag{7}$$

$$(1 + adX^* + 2aX^*M_2)^2 < (q - X^*)(h + aX^*)$$

then E^* is globally asymptotically stable.

Proof. We consider the following positive definite function about E^*

$$L(X,Y,Z) = \left(X - X^* - X^* ln\frac{X}{X^*}\right) + \frac{1}{2}(Y - Y^*)^2 + \frac{1}{2}(Z - Z^*)^2$$

Differentiating both sides with respect to t along the solution of (3), we get (after some simple calculations):

$$\begin{aligned}
\frac{dL}{dt} = &- (X - X^*)^2 + (Y - 1)(X - X^*)(Y - Y^*) - (q - X^*)(Y - Y^*)^2 \\
&+ (aY(d + Y) - aZ)(X - X^*)(Z - Z^*) - (aX^* + h)(Z - Z^*)^2 \\
&+ (1 + adX^* + aX^*(Y + Y^*))(Y - Y^*)(Z - Z^*) \\
\\
\leq &- (X - X^*)^2 + (M_2 - 1)(X - X^*)(Y - Y^*) - (q - X^*)(Y - Y^*)^2 \\
&- (aX^* + h)(Z - Z^*)^2 + (aM_2(d + M_2) - am_3)(X - X^*)(Z - Z^*) \\
&+ (1 + adX^* + 2aM_2X^*)(Y - Y^*)(Z - Z^*) \\
\\
= &-a_{11}(X - X^*)^2 - a_{22}(Y - Y^*)^2 - a_{33}(Z - Z^*)^2 + a_{12}(X - X^*)(Y - Y^*) \\
&+ a_{13}(X - X^*)(Z - Z^*) + a_{23}(Y - Y^*)(Z - Z^*)
\end{aligned}$$

where

$$a_{11} = 1, \quad a_{22} = q - X^*, \quad a_{33} = aX^* + h, \quad a_{12} = M_2 - 1,$$

$$a_{13} = \{aM_2(d + M_2) - am_3\}, \quad a_{23} = 1 + adX^* + 2aM_2X^*$$

Now sufficient conditions for $\frac{dL}{dt}$ to be negative definite are

$$a_{12}^2 - a_{11}a_{22} < 0, \quad a_{13}^2 - a_{11}a_{33} < 0, \quad a_{22}^2 - a_{22}a_{33} < 0, \tag{8}$$

i.e.,

$$(M_2 - 1)^2 < (q - X^*)$$

$$\{aM_2(d + M_2) - am_3\}^2 < (aX^* + h)$$

$$(1 + adX^* + 2aX^*M_2)^2 < (q - X^*)(h + aX^*).$$

Hence L is a Lyapunov function [28–30]. This competes the theorem.

4   Model with Double Delays

In this section we have discussed the model with double discrete time delays τ_1 and τ_2, where τ_1 represents the activation period or reaction time of the toxicant in the population biomass and τ_2 represents the activation period or reaction time of the toxicant in the population from the environment. The model is as follows:

$$\frac{dX}{dt} = X\left(p - Y(t-\tau_1) - X\right)$$

$$\frac{dY}{dt} = Z(t-\tau_2) - (q-X)Y \qquad (9a)$$

$$\frac{dZ}{dt} = -aXZ + a(d+Y)XY - hZ + Q$$

$$X(0) \geq 0; \quad Y(\theta_1) \geq 0, \quad \theta_1 \in [-\tau_1, 0]; \quad Z(\theta_2) \geq 0, \quad \theta_2 \in [-\tau_2, 0]. \qquad (9b)$$

4.1   Estimation of the length of delay to preserve stability

Now, we shall estimate the length of the delay which preserves the stability of the system (9). The corresponding linearized system of (9) about the interior equilibrium $E^*(X^*, Y^*, Z^*)$ is given by:

$$\frac{dx_1}{dt} = a_{11}x_1 + a_{12}y_1(t-\tau_1)$$

$$\frac{dy_1}{dt} = a_{21}x_1 + a_{22}y_1 + b_{23}z_1(t-\tau_2) \qquad (10)$$

$$\frac{dz_1}{dt} = a_{31}x_1 + a_{32}y_1 + a_{33}z_1$$

where

$$X(t) = x_1(t) + X^*, \quad Y(t) = y_1(t) + Y^*, \quad Z(t) = z_1(t) + Z^*.$$

Taking Laplace transform of the system (10) we get,

$$(s - a_{11})\bar{x}_1(s) = a_{12}e^{-s\tau_1}\bar{y}_1(s) + a_{12}e^{-s\tau_1}k_1(s) + x_1(0)$$

$$(s - a_{22})\bar{y}_1(s) = a_{21}\bar{x}_1(s) + b_{23}e^{-s\tau_2}\bar{z}_1(s) + b_{23}e^{-s\tau_2}k_1(s) + y_1(0) \qquad (11)$$

$$(s - a_{33})\bar{z}_1(s) = a_{31}\bar{x}_1(s) + a_{32}\bar{y}_1(s) + z_1(0)$$

where

$$k_1(s) = \int_{-\tau_1}^{0} e^{-st}y_1(t)\,dt,$$

$$k_2(s) = \int_{-\tau_2}^{0} e^{-st}z_1(t)\,dt$$

and $\bar{x}_1(s), \bar{y}_1(s), \bar{z}_1(s)$ are the Laplace transforms of $x_1(t), y_1(t), z_1(t)$ respectively.

Now, using "Nyquist theorem" [21, 31], it can be shown that the conditions for local asymptotic stability of $E^*(X^*, Y^*, Z^*)$ are given by [32]

$$ImH(i\eta_0) > 0, \tag{12}$$

$$ReH(i\eta_0) = 0, \tag{13}$$

where

$$H(s) = s^3 + D_1 s^2 + D_2 s + D_3 + D_4 s e^{-s\tau_1} + D_5 s e^{-s\tau_2} + D_6 e^{-s\tau_1} + D_7 e^{-s\tau_2} + D_8 e^{-s(\tau_1+\tau_2)} \tag{14}$$

and

$$\begin{aligned}
D_1 &= aX^* + h + q, \\
D_2 &= aqX^* + hq + qX^* - (X^*)^2, \\
D_3 &= aq(X^*)^2 + qhX^* - a(X^*)^3 - h(X^*)^2, \\
D_4 &= X^*Y^*, \\
D_5 &= -a(d + Y^*)X^* - aX^*Y^*, \\
D_6 &= X^*Y^*(aX^* + h), \\
D_7 &= -X^*[a(d + Y^*)X^* - aX^*Y^*], \\
D_8 &= X^*[-aZ^* + a(d + Y^*)Y^*]
\end{aligned}$$

and η_0 is the smallest positive root of equation (13).

Nyquist criterion [21, 31] implies that if η_0 is a solution of equation (13), then

$$\begin{aligned}
&-D_1\eta_0^2 + D_3 + D_4\eta_0\sin(\eta_0\tau_1) + D_5\eta_0\sin(\eta_0\tau_2) + D_6\cos(\eta_0\tau_1) \\
&+ D_7\cos(\eta_0\tau_2) + D_8\cos\{\eta_0(\tau_1+\tau_2)\} = 0
\end{aligned} \tag{15}$$

Next, we wish to find an upper bound for η (say, η_+) independent of τ_1, τ_2 such that (12) and (13) hold $\forall \eta, 0 \leq \eta \leq \eta_+$ and hence in particular for η_0 when $\eta_0 \leq \eta_+$.

Maximizing (15) with $|\sin(\eta_0\tau_1)| \leq 1, |\sin(\eta_0\tau_2)| \leq 1, |\cos(\eta_0\tau_1)| \leq 1,$ $|\cos(\eta_0\tau_2)| \leq 1, |\cos\{\eta_0(\tau_1+\tau_2)\}| \leq 1$, we get,

$$D_1\eta_0^2 - (|D_4| + |D_5|)\eta_0 - (|D_3| + |D_6| + |D_7| + |D_8|) = 0 \tag{16}$$

If η_+ is a positive root of (16), then we obtain

$$\eta_+ = \frac{1}{2D_1}[(|D_4| + |D_5|) + \{(|D_4| + |D_5|)^2 + 4D_1(|D_3| + |D_6| + |D_7| + |D_8|)\}^{\frac{1}{2}}] \tag{17}$$

Then clearly, $\eta_+ \geq \eta_0$.

Equation (12) indicates that the following inequality should hold:

$$\varphi(\tau_1, \tau_2, \eta_0) > \psi(\tau_1, \tau_2, \eta_0),$$

where

$$\varphi(\tau_1, \tau_2, \eta_0) = -\eta_0[\eta_0^2 - D_2 - D_4\cos(\eta_0\tau_1) - D_5\cos(\eta_0\tau_2)] \qquad (18)$$

and

$$\psi(\tau_1, \tau_2, \eta_0) = D_6\sin(\eta_0\tau_1) + D_7\sin(\eta_0\tau_2) + D_8\sin\{\eta_0(\tau_1 + \tau_2)\}]. \qquad (19)$$

If we can find $\tilde{\varphi}(\tau_1, \tau_2) > \tilde{\psi}(\tau_1, \tau_2)$ such that

$$\frac{\varphi(\tau_1, \tau_2, \eta_0)}{(\tau_1 + \tau_2)\eta_0} \geq \tilde{\varphi}(\tau_1, \tau_2) > \tilde{\psi}(\tau_1, \tau_2) \geq \frac{\psi(\tau_1, \tau_2, \eta_0)}{(\tau_1 + \tau_2)\eta_0}, \qquad (20)$$

where $0 < \eta_0 < \eta_+$, then the Nyquist criterion holds. By equation (20) we can estimate the values of τ_1 and τ_2.

$$\frac{\psi(\tau_1, \tau_2, \eta_0)}{(\tau_1 + \tau_2)\eta_0} = \frac{D_6\sin(\eta_0\tau_1) + D_7\sin(\eta_0\tau_2) + D_8\sin\{\eta_0(\tau_1 + \tau_2)\}}{(\tau_1 + \tau_2)\eta_0}$$

$$\leq |D_6| + |D_7| + |D_8|$$

We choose,

$$\tilde{\psi}(\tau_1, \tau_2) = |D_6| + |D_7| + |D_8|. \qquad (21)$$

Now,

$$\frac{\varphi(\tau_1, \tau_2, \eta_0)}{(\tau_1 + \tau_2)\eta_0} = \frac{\eta_0\{D_2 + D_4\cos(\eta_0\tau_1) + D_5\cos(\eta_0\tau_2) - \eta_0^2\}}{(\tau_1 + \tau_2)\eta_0}$$

$$\geq \frac{\{D_2 + D_4\cos(\eta_+\tau_1) + |D_5| - \eta_+^2\}}{(\tau_1 + \tau_2)}$$

We can choose,

$$\tilde{\varphi}(\tau_1, \tau_2) = \frac{\{D_2 + D_4\cos(\eta_+\tau_1) + |D_5| - \eta_+^2\}}{(\tau_1 + \tau_2)}. \qquad (22)$$

From (20) we get,

$$\tilde{\varphi}(\tau_1, \tau_2) = \frac{\{D_2 + D_4\cos(\eta_+\tau_1) + |D_5| - \eta_+^2\}}{(\tau_1 + \tau_2)} > |D_6| + |D_7| + |D_8|$$

$$= \tilde{\psi}(\tau_1, \tau_2).$$

Thus

$$\tau_1 + \tau_2 < \frac{D_2 + D_4\cos(\eta_+\tau_1) + |D_5| - \eta_+^2}{|D_6| + |D_7| + |D_8|}$$
$$\leq \frac{|D_2| + |D_4| + |D_5| - \eta_+^2}{|D_6| + |D_7| + |D_8|}. \tag{23}$$

So, we come to the following theorem:

Theorem 6. *The delayed model (9) will be locally asymptotically stable at $E^*(X^*, Y^*, Z^*)$ with the conditions (6) if*

$$|D_2| + |D_4| + |D_5| > \eta_+^2$$

and

$$0 < \tau_1 + \tau_2 < \frac{|D_2| + |D_4| + |D_5| - \eta_+^2}{|D_6| + |D_7| + |D_8|}.$$

4.2   Analysis of existence of Hopf-bifurcation

Now, we will study the existence of Hopf-bifurcation of system (9) by choosing one of the delays as a bifurcation parameter, i.e., taking τ_2 as a bifurcation parameter. The characteristic equation corresponding to system (9) is

$$\lambda^3 + D_1\lambda^2 + D_2\lambda + D_3 + D_4\lambda e^{-\lambda\tau_1} + D_5\lambda e^{-\lambda\tau_2} + D_6 e^{-\lambda\tau_1} + D_7 e^{-\lambda\tau_2} + D_8 e^{-\lambda(\tau_1+\tau_2)} = 0. \tag{24}$$

Hence, $\lambda = \pm i\eta$ is a simple root of (24) at $\tau_2 = \tau_2^c$. By separating real and imaginary parts, we obtain

$$\begin{aligned} &-D_1\eta^2 + D_3 + D_4\eta\sin(\eta\tau_1) + D_5\eta\sin(\eta\tau_2) \\ &+D_6\cos(\eta\tau_1) + D_7\cos(\eta\tau_2) + D_8\cos\{\eta(\tau_1+\tau_2)\} = 0, \end{aligned} \tag{25}$$

and

$$\begin{aligned} &-\eta^3 + D_2\eta + D_4\eta\cos(\eta\tau_1) + D_5\eta\cos(\eta\tau_2) \\ &-D_6\sin(\eta\tau_1) - D_7\sin(\eta\tau_2) + D_8\sin\{\eta(\tau_1+\tau_2)\} = 0. \end{aligned} \tag{26}$$

Differentiating equation (24) implicitly with respect to τ_2, we obtain

$$\begin{aligned} \left(\frac{d\lambda}{d\tau_2}\right)^{-1} &= \frac{D_5}{\lambda(D_5\lambda + D_7 + D_8 e^{-\lambda\tau_1})} + \frac{D_8\tau_1 e^{-\lambda\tau_1}}{\lambda(D_5\lambda + D_7 + D_8 e^{-\lambda\tau_1})} \\ &+ \frac{3\lambda^2 + 2D_1\lambda + D_2}{\lambda e^{-\lambda\tau_2}(D_5\lambda + D_7 + D_8 e^{-\lambda\tau_1})} + \frac{e^{-\lambda\tau_1}(D_4 - D_4\lambda\tau_1 - D_6\tau_1)}{\lambda e^{-\lambda\tau_2}(D_5\lambda + D_7 + D_8 e^{-\lambda\tau_1})} - \frac{\tau_2}{\lambda} \end{aligned}$$

Now,

$$Re\left(\frac{d\lambda}{d\tau_2}\right)^{-1}_{\lambda=i\eta_0} = \frac{1}{M^2 + N^2}(MP + NR), \tag{27}$$

where,

$$M = -D_5\eta_0^2 + D_8\eta_0\sin(\eta_0\tau_1),$$
$$N = D_7\eta_0 + D_8\eta_0\cos(\eta_0\tau_1),$$
$$P = D_5 - D_8\tau_1\cos(\eta_0\tau_1) + (-3\eta_0^2 + D_2)\cos(\eta_0\tau_2) - 2D_1\eta_0\sin(\eta_0\tau_2)$$
$$+\{(D_4 - D_6\tau_1)\cos(\eta_0\tau_1) - \eta_0\tau_1 D_4\sin(\eta_0\tau_1)\}\cos(\eta_0\tau_2)$$
$$+\{(D_4 - D_6\tau_1)\sin(\eta_0\tau_1) + \eta_0\tau_1 D_4\cos(\eta_0\tau_1)\}\sin(\eta_0\tau_2)$$
$$R = D_8\tau_1\sin(\eta_0\tau_1) + (-3\eta_0^2 + D_2)\sin(\eta_0\tau_2) + 2D_1\eta_0\cos(\eta_0\tau_2)$$
$$+\{(D_4 - D_6\tau_1)\cos(\eta_0\tau_1) - \eta_0\tau_1 D_4\sin(\eta_0\tau_1)\}\sin(\eta_0\tau_2)$$
$$-\{(D_4 - D_6\tau_1)\sin(\eta_0\tau_1) + \eta_0\tau_1 D_4\sin(\eta_0\tau_1)\}\cos(\eta_0\tau_2).$$

Cleary the sign of $M^2 + N^2$ in (27) is positive. Because of the coherence of the sign of $MP + NR$ and $Re\left(\frac{d\lambda}{d\tau_2}\right)^{-1}_{\lambda=i\eta_0}$, we get, if $MP + NR > 0$ then $Re\left(\frac{d\lambda}{d\tau_2}\right)^{-1}_{\lambda=i\eta_0}$ > 0.

Therefore, transversility condition holds and Hopf-bifurcation occurs at $\tau_2 = \tau_2^c$, provided $(MP + NR) > 0$.

Summarizing the above analysis we get the following result.

Theorem 7. *If there exists $\tau_2 = \tau_2^c$ such that (25) and (26) hold and $Re\left(\frac{d\lambda}{d\tau_2}\right)^{-1}_{\lambda=i\eta_0}$* > 0, *that is, $(MP + NR) > 0$, then a Hopf-bifurcation occurs at $E^*(X^*,Y^*,Z^*)$ as τ_2 passes through τ_2^c.*

5   Numerical Simulations

In this section we present computer simulation of some important analytical results discussed in the previous sections. We take the parameters of the system as given in Table 1. Using those values under zero exogeneous input, i.e., $\gamma = 0$ it is observed that the equilibrium $E_0(0,0,0)$ becomes unstable and interior equilibrium $E^*(X^*,Y^*,Z^*)$ is not feasible. Only the axial equilibrium $E_1(p,0,0) = (2.28,0,0)$ is locally asymptotically stable, which is shown in Fig. 1(a). Therefore, Theorem 2 is verified. When $\gamma = 1.25$ (i.e., non-zero exogeneous input) using the parameter values given in Table 2, it is observed that the equilibrium $E_2 = (0,4.28082,12.5)$ exists and locally asymptotically stable with $hpq < \gamma$, but the interior equilibrium E^* does not exist, which is shown in Fig. 1(b). On the other hand, when $\gamma = 1.25$ using the parameter values given in Table 1, it is found that $hpq > \gamma, a(q - p - d) > h$, and $E_2 = (0,2.14041,6.25)$ becomes unstable and $E^*(X^*,Y^*,Z^*) = (0.43227,1.84773, 4.59665)$ exists, which is shown in Fig. 2(a). Therefore, Theorem 3 is verified.

Using the parameter values given in Table 1 with $\gamma = 1.25$, it is observed that the conditions of Theorem 4 are satisfied as $A_1 = 4.9096 > 0, A_3 = 0.121507 > 0, A_1A_2 - A_3 = 0.388821 > 0$. Therefore by Theorem 4, E^* is locally asymptotically stable. Fig. 2(a) shows that X, Y and Z approach to their steady-state values X^*, Y^* and Z^* respectively in finite time. The XY-plane, YZ-plane, XZ-plane and XYZ-plane projections of the solution are shown in Figs.2(b)-2(e) respectively. Here the conditions of Theorem 5 are also satisfied and consequently E^* is globally asymptotically stable. The phase diagram is shown in Fig. 2(f).

Parameter	Values
p	2.28
q	2.92
a	4.14
d	0.54
h	0.2

Table 1

Parameter	Values
p	2.28
q	2.92
a	4.14
d	0.54
h	0.1

Table 2

Fig. 1(a)

Fig. 1(b)

Fig. 1.(a) Time series plot of X, Y, Z for zero exogeneous input ($\gamma = 0$) with $X(0) = 1, Y(0) = 2, Z(0) = 0.2$, parameter values are given in Table 1, (b) Time series plot of X, Y, Z for non-zero exogeneous input ($\gamma = 1.25$) and $hpq < \gamma$ with $X(0) = 3, Y(0) = 5, Z(0) = 8$, parameter values are given in Table 2.

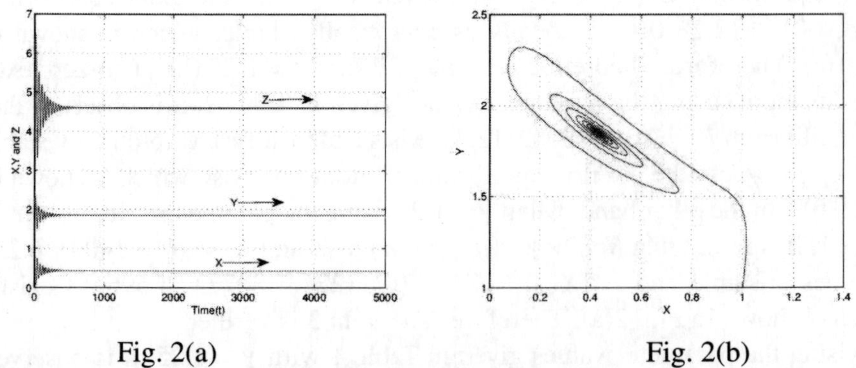

Fig. 2(a)

Fig. 2(b)

Fig. 2.(a) Time series plot of X, Y, Z for non-zero exogeneous input ($\gamma = 1.25$) and $hpq > \gamma$ with $X(0) = 1, Y(0) = 1, Z(0) = 5$, parameter values are given in Table 1, (b) XY-plane projection of the solution for non-zero exogeneous input

($\gamma = 1.25$) with $X(0) = 1, Y(0) = 1, Z(0) = 5$, parameter values are given in Table 1.

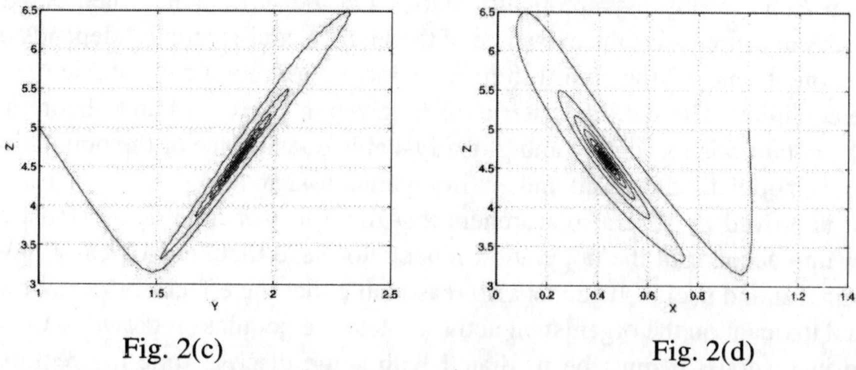

Fig. 2(c)   Fig. 2(d)

Fig. 2.(c) YZ-plane projection of the solution for non-zero exogeneous input ($\gamma = 1.25$) with $X(0) = 1, Y(0) = 1, Z(0) = 5$, parameter values are given in Table 1, (d) XZ-plane projection of the solution for non-zero exogeneous input ($\gamma = 1.25$) with $X(0) = 1, Y(0) = 1, Z(0) = 5$, parameter values are given in Table 1.

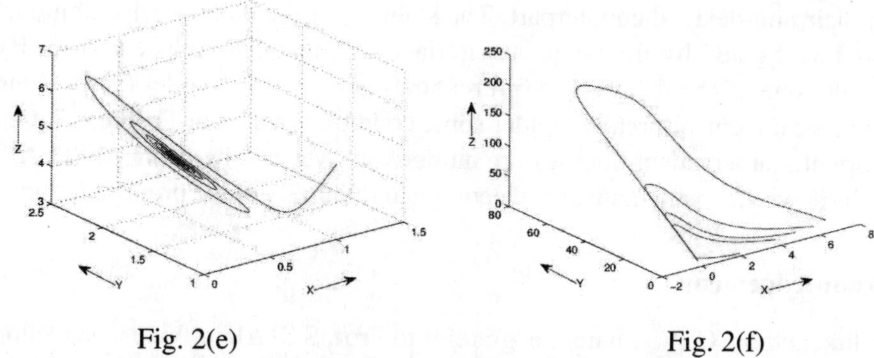

Fig. 2(e)   Fig. 2(f)

Fig. 2.(e) XYZ-plane projection of the solution for non-zero exogeneous input ($\gamma = 1.25$) with $X(0) = 1, Y(0) = 1, Z(0) = 5$, parameter values are given in Table 1, (f) Global stability of $E^*(X^*, Y^*, Z^*)$ for non-zero exogeneous input ($\gamma = 1.25$) with $(X(0), Y(0), Z(0)) = (2,1,3), (4,3,5), (6,4,8), (5,3,8), (7,8,4)$, parameter values are given in Table 1.

6   Discussions and Conclusions

In this paper, we have developed a single species population model in the polluted population by means of ordinary differential equations in terms of their concentrations. Boundedness, positivity, stability analysis of the model (3) with (4) at various equilibrium points are discussed thoroughly, which implies that the system is ecologically well-behaved. The analysis of the existence of various equilibrium points under zero exogeneous toxicant input and non-zero exogeneous toxicant input lead us to Theorem 2 and Theorem 3 respectively.

Theorem 2 states that under zero exogeneous input the trivial equilibrium E_0 is unstable, the interior equilibrium E^* is not feasible and only the axial equilibrium E_1 is locally asymptotically stable. On the other hand when exogeneous input is non-zero, the existence of the interior equilibrium E^* depends on some conditions and the equilibrium E_2 is locally asymptotically stable under some condition. The stability criteria for E^* given in Theorem 4 and Theorem 5 are the conditions for locally and globally stable coexistence of the population biomass, population toxicant and environmental toxicant.

It is proved by several researchers that the effect of time delays must be taken into account in the population models to make them ecologically more meaningful and useful. It seems also reasonable that the effect of the environmental toxicant on the organismal activities and the population growth will not be instantaneous. It must be mediated with some discrete time lag required for incubation. From this point of view we have formulated double discrete time delayed model given in (9), where the delays may be looked upon as the reaction time of the environment toxicant on the population biomass and the population toxicant.

The analysis of the delayed model (9) shows more complicated behaviour than their non-delayed counterpart. The stability of the double delayed model (9) is investigated by the Nyquist criteria which leads us to Theorem 6. By choosing one of the delays as a bifurcation parameter the model (9) is found to undergo a Hopf-bifurcation under some conditions stated in Theorem 7. Our important mathematical findings are numerically verified by using MATLAB, which shows the mathematical and ecological reliability of the proposed model.

Acknowledgement

The first author (G.P. Samanta) is grateful to Prof. S.S. Alam, Vice Chancellor, Aliah University for his help and encouragement. He is also thankful to Dr. J. Mukhopadhyay, Dr. Nurul Huda Gazi and Dr. Sk. Md. Abu Nayeem, Aliah University for their cooperation.

References

1. T. G. Hallam, C. E. Clark, G. S. Jordan, Effects of toxicants on populations: A qualitative approach II. First order kinetics, J. Math. Biol., 18 (1983), 25–37.
2. T. G. Hallam, C. E. Clark, R. R. Lassiter, Effects of toxicants on populations: A qualitative approach I. Equilibrium environmental expose, Ecol. Model., 18 (1983), 291–304.
3. T.G. Hallam, J.T. De Lena, Effects of toxicants on populations: A qualitative approach III. Environmental and food chain pathways, J. Theor. Biol., 109 (1984), 411–429.
4. Z. He, Z. Ma, On the effects of polution and catch to a logistic population, J. Biomath., 14 (1999), 281–287.
5. Z. He, Z. Ma, The effect of toxicant and harvest on the logistic model, J. Biomath., 12 (1997), 230–237.

6. Z. Ma, The Study of Biology Model, He Fei, Anhui Education Press, 1996.

7. Z. Ma, B. Song, T.G. Hallam, The threshold between persistence and extinction of a population in a polluted environment, Bulletin of Mathematical Biology, 51 (1989), 311–323.

8. B. Buonomo, A.D. Liddo, I. Sgura, A diffusive-connective model for the dynamics of population-toxicant intentions: Some analytical and numerical results, Math. Biosci., 157 (1999), 37–64.

9. J.T. De Lena, T.G. Hallam, Effects of toxicants on population: A qualitative approach IV, Ecol. Model., 35 (1987), 249–273.

10. H.I. Freedman, J.B. Shukla, Models for the effect of toxicant in single species and predator-prey system, J. Math. Biol., 30 (1991), 15–30.

11. M. Ghosh, P. Chandra, P. Sinha, A mathematical model to study the effect of toxicant chemicals on a prey-predator type fishery, J. Biol. Syst., 10 (2002), 97–105.

12. J. He, K. Wang, The survival analysis for a population in a polluted environment, Nonlinear Analysis: Real World Applications, 3 (2009), 1555–1571.

13. Z. Li, Z. Shuai, K. Wang, Persistence and extinction of single population in a polluted environment, Electron. J. Diff. Eqn., 108 (2004), 1–5.

14. G.P. Samanta, A. Maity, Dynamical model of a single species model in a polluted environment, J. Appl. Math. Comput., 16 (2004), 231–242.

15. J. Wang, K. Wang, Analysis of a single species with diffusion in a polluted environment, Electron, J. Diff. Eqn., 112 (2006), 1–11.

16. A. Buonomo, The periodic solution of Van der Pol's equation, SIAM J. Appl. Math., 59 (1998), 156–171.

17. C. Celik, The stability and Hopf bifurcation for a predator-prey system with time delay, Chaos, Solitons and Fractals, 37 (2008), 87–99.

18. Y. Chen, J. Yu, C. Sun, Stability and hopf bifurcation analysis in a three-level food chain system with delay, Chaos, Solitons and Fractals, 31 (2007), 683–694.

19. K. Gopalsamy, Stability and Oscillations in Delay Differential Equations of Population Dynamics, Kluwer Academic, Dordrecht, 1992.

20. Y. Kuang, Delay Differential Equations with Applications in Population Dynamics, Academic Press, New York, 1993.

21. X. Liao, Hopf and resonant codimension two bifurcation in Van Dar Pol equation with two time delays, Chaos, Solutons and Fractals, 23 (2005), 857–871.

22. W. Liu, Criterion of Hopf bifurcations without using eigen values, J. Math. Anal. Appl., 182 (1994), 250–256.

23. A. Maiti, A. K. Pal, G. P. Samanta, Effect of time delay on a food chain model, Appl. Math. Comput., 200 (2008), 189–203.

24. A. Maiti, A. K. Pal, G. P. Samanta, Usefulness of biocontrol of pests in tea: a mathematical model, Math. Model. Nat. Phenom., 3 (2008), 96–113.

25. Y. Song, M. Han, Y. Peng, Stability and Hopf bifurcation in competitive Lotka-Volterra system with two delays, Chaos, Solitons and Fractals, 22 (2004), 1139–1148.

26. Z. H. Wang, H. H. Hu, Stability of linear time variant dynamic systems with multiple time delays, Acta Mechanica Sinica, 14 (1998), 274–282.

27. R. Xua, Q. Gan, Z. Ma, Stability and bifurcation analysis on a ratio dependent predator-prey model with time delay, J. Comput. Appl. Math., 230 (2009), 187–203.

28. H. I. Freedman, Deterministic Mathematical Models in Populations Ecology, Marcel Dekker, New York, 1980.

29. M. Kot, Elements of Mathematical Ecology, Cambridge University Press, Cambridge, 2001.

30. J. D. Murray, Mathematical Biology, Springer-Verlag, New York, 1993.

31. H. Freedman, V. S. H. Rao, The trade-off between mutual interference and time lags in predator-prey systems, Bull. Math. Biol., 45 (1983), 991–1004.

32. L. H. Erbe, V. S. H. Rao, H. Freedman, Three species food chain models with mutual interference and time delays, Math. Biosci., 80 (1986), 57–80.

Endomorphism Rings of Reduced Modules

Ardeline Mary Buhphang[*]

Department of Mathematics, North Eastern Hill University, Permanent Campus,
Shillong – 793 022, Meghalaya, India.

Abstract. In this paper we study the relationships between a reduced module and its endomorphism ring. We show that for a cyclic module M over a left duo ring R, $_RM$ is reduced if and only if its ring of endomorphisms $End_R(M)$ is a reduced ring if and only if M is reduced as a module over $End_R(M)$.

Key words: reduced ring, reduced module, semi-simple ring, von Neumann regular ring, reduced ring.

1   Introduction

All the rings are associative with $1 \neq 0$ and all the modules are unitary modules. A ring is *reduced* if it has no non zero nilpotent elements. A ring R is *reversible* (see [1]) if whenever elements $a, b \in R$ satisfy $ab = 0$, we have $ba = 0$. Every reduced ring is reversible but the converse is not true as commutative non reduced rings are reversible. A ring R is *von Neumann regular* (or simply *regular*) if for every $a \in R$, there exists $b \in R$ such that $a = aba$. R is *strongly regular* if for every $a \in R$, there exists $b \in R$ such that $a = ba^2$. Strongly regular rings are regular. A regular ring is strongly regular if and only if it is reduced. A ring R is *left duo* if every left ideal of R is two-sided. A left R-module M is *reduced* if whenever $a \in R$, $m \in M$ satisfy $a^2 m = 0$, we have $aRm = 0$. Equivalently, (see [2]) whenever $a \in R$, $m \in M$ satisfy $am = 0$, we have, $Rm \cap aM = 0$. A left R-module M is *semisimple* if every submodule of M is a direct summand of M. Following [3], a module $_RM$ is *regular* if for every $m \in M$, there exists $f \in S$ such that $(mf)m = m$. It is to be noted here that module homomorphisms are written on the side opposite to that of scalars.

The notation $_RM$ will stand for the left R-module M and S for the endomorphism ring $End_R(M)$. Recall that if a module $_RM$ is faithful, then R is a submodule of direct product of copies of M. Using this fact and the fact that for any module $_RM$, the module M_S is faithful, Rege and Buhphang [4] showed in Remark 2.5 that if M_S is a reduced module then S is a reduced ring. Motivated by this remark, we attempt in this paper to find out relations between the ring S, the module $_RM$ and the module M_S in terms of the reduced conditions of rings and modules.

---

[*] Email: ardeline17@gmail.com

2 Main Results

We begin with the following two lemmas.

Lemma 1. *([5], Proposition 4.27) If $_RM$ is a semi-simple module, then S is von Neumann regular.*

Lemma 2. *([4], Theorem 2.16) A ring R is strongly regular if and only if every left R-module is reduced.*

Proposition 1. *If $_RM$ is a semi-simple module and if S is a reduced ring, then the module M_S is reduced.*

Proof. By Lemma 1, since $_RM$ is semi-simple, S is a regular ring. Therefore S is a strongly regular ring and so by Lemma 2, M_S is a reduced module.

If $_RM$ is a reduced module then S may not be a reduced ring. For example the $\mathbb{Z}$-module $\mathbb{Z} \times \mathbb{Z}$ is reduced, but $S = M_2(\mathbb{Z})$ is not a reduced ring. However we have the results:

Proposition 2. *If M is a cyclic and reduced left R-module, then the ring S is reduced.*

Proof. Let $M = Rm$ for some $m \in M$ and let $f \in S$ such that $f^2 = 0$. Then $mf = am$ for some $a \in R$ and so $0 = mf^2 = (mf)f = (am)f = a^2m$. As $_RM$ is reduced, we get $mf = am = 0$ and hence $f = 0$.

Proposition 3. *If M is a cyclic, regular module over a reduced ring R then S is reduced.*

Proof. By Corollary 1.5 of [3], M is isomorphic to a left ideal Re of R generated by an idempotent e. Again by Proposition 4.6 of [6], $End_R(Re)$ is isomorphic to eRe which is a reduced ring since R itself is reduced. So we get that S is a reduced ring.

Let P be a property on a ring R. We say R is *completely P* if every factor ring of R has P.

Proposition 4. *If R is a reduced and completely reversible ring, then for every left ideal A of R, $End_R(A)$ is a reduced ring.*

Proof. Let $f \in End_R(A)$ such that $f^2 = 0$. Then for all $a \in A$, we have,

$$0 = (a^2)f^2 = ((a^2)f)f = (a(af))f$$

which means that $a(af) \in ker(f)$, equivalently, $a(af) = 0$ in the factor ring $R/ker(f)$. As $R/ker(f)$ is reversible, we get, $(af)a = 0$ in $R/ker(f)$ and so $0 = ((af)a)f = (af)(af) = (af)^2$. Since R is reduced, we get $af = 0$ and hence $f = 0$.

Corollary 1. *If R is a completely reduced ring, then for every left ideal A of R, $End_R(A)$ is a reduced ring.*

It may not be true that if $_RM$ is a reduced module then M_S would be a reduced module. For example the $\mathbb{Z}$-module $\mathbb{Z} \times \mathbb{Z}$ is reduced, but as $S = M_2(\mathbb{Z})$, taking $m = (1,0)$, $f = \begin{pmatrix} 0 & 1 \\ 0 & 0 \end{pmatrix}$, we get $mf^2 = 0$, but $mf \neq 0$. However,

Proposition 5. *If $_RM$ is a cyclic module which is reduced, then M_S is also a reduced module.*

Proof. Suppose that $M = Rm_0$ and let $m \in M$, $f \in S$ such that $mf = 0$. Then $m_0 f = am_0$ and $m = bm_0$ for some $a, b \in R$. So, $0 = (mf) = (bm_0)f = b(am_0)$. As $_RM$ is reduced, we have, $Ram_0 \cap bM = 0$. Now let $y \in mS \cap Mf$. Then $y = mg = nf$ for some $g \in S$, $n \in M$. So $y = mg = b(m_0g)$ and $y = nf = (cm_0)f = cam_0$ for some $a \in R$. Thus $y \in Ram_0 \cap bM = 0$ and so M_S is reduced.

Theorem 1. *Let R be a left duo ring. The following conditions are equivalent for a cyclic left R-module M.*

1. *$_RM$ is a reduced module.*
2. *S is a reduced ring.*
3. *M_S is a reduced module.*

Proof. The implication $(1) \Rightarrow (2)$ is by Proposition 4. The implication $(3) \Rightarrow (2)$ is by Remark 2.5 of [4]. Since M is a cyclic left R-module, we have $M \cong R/B$ for some left ideal (which is two-sided in this case since R is a left duo ring) B of R. For $(2) \Rightarrow (3)$, it can be shown that R/B is an S-submodule of S and therefore since S is reduced as a module over itself, R/B is reduced over S. We finally show that $(2) \Rightarrow (1)$. Let $a, x \in R$ such that $a^2\bar{x} = 0$. Consider $f, g \in S$ such that $f(\bar{r}) = \overline{rax}$, $g(\bar{r}) = \overline{ra}$, for all $r \in R$. Both f and g are well-defined as R is a left duo ring. Then $f \circ g = 0$. As S is reduced, $g \circ f = 0$ and therefore we get $f \circ f = 0$. Since S is a reduced ring, $f = 0$ and so $a\bar{x} = 0$. Again for any $r \in R$, letting $(\bar{t}f_r) = \overline{tr}$, $\forall \bar{t} \in R/B$, we get $f_x \circ f_a = 0$. So as R is reduced, for all $r \in R$, we have $f_a \circ f_r \circ f_x = 0$ and hence

$$ar\bar{x} = ar(\bar{1}f_x) = a(\bar{r}f_x) = a(\bar{1}f_rf_x) = (\bar{a}f_s)f_x = (\bar{1})f_af_rf_x = 0.$$

Thus $_RM$ is reduced.

We end this paper by a result on the dual module. Recall that for a left R-module M, the dual module $Hom_R(M,R)$ denoted by M^* is a right R-module via $(m)(f.a) = (mf)a$. Following these notations, we have

Proposition 6. *For any left module M over a reduced ring, the right module M_R^* is always reduced.*

Proof. Let $a \in R$, $f \in M^*$ such that $fa^2 = 0$. Then for all $m \in M$, $(mf)a^2 = 0$. As reduced rings are reversible, this yields $a(mf)a = 0$ hence $(a(mf))^2 = 0$. As R is reduced, $a(mf) = 0$ which (again by reversibility of R) gives $(mf)ra = 0$, for all $r \in R$. Thus M_R^* is a reduced module.

References

1. P.M. Cohn, Reversible rings, Bull. London Math. Soc., 31 (1999), 641–648.
2. T.K. Lee and Y. Zhou, Reduced modules, Lecture notes in Pure and Applied Math. 236, Marcel Dekker, New York, 2004.
3. J. Zelmanowitz, Regular modules, Trans. Amer. Math. Soc., 163 (1972), 341–355.
4. M.B. Rege and A.M. Buhphang, On reduced modules and rings, Intl. Elec. Journal of Alg., 3 (2008), 58–74.
5. T.Y. Lam, A First Course in Noncommutative Rings, Springer-Verlag, New York, 1991.
6. F.W. Anderson and K.R. Fuller, Rings and Categories of Modules, Springer-Verlag, New York, 1973.

A Note on Convolution Properties

N. Shilpa[†,*], L. Dileep[‡,**] and S. Latha[†,***]

[†]Department of Mathematics, Yuvaraja's College, University of Mysore,
Mysore – 570 005, India.
[‡]Department of Mathematics, Vidyavardhaka College of Engineering,
Gokolum 3rd stage, Mysore, India.

Abstract. In this note, we define the subclasses $\Omega_\lambda^\delta(A,B)$ and $\Omega^\delta(A,B)$ of analytic functions using the fractional derivative Ω^δ. Further characterization theorem and convolution properties for the above mentioned classes are determined.

Key words: Fractional derivative, convolution condition, Janowski class.

1   Introduction

Let $\mathscr{A}$ denote the class of analytic functions in the open unit disc
$\mathscr{U} = \{z; |z| < 1\}$ of the form

$$f(z) = z + \sum_{n=2}^{\infty} a_n z^n. \tag{1}$$

For $-1 \le B < A \le 1$, let $\mathscr{P}(A,B)$ [1] denote the class of functions which are of the form

$$p(z) = \frac{1+A\omega(z)}{1+B\omega(z)},$$

where ω is a bounded analytic function satisfying the conditions
$\omega(0) = 0 \quad \text{and} |\omega(z)| < 1.$

The Fractional derivative of order δ of an analytic function f is defined by

$$D_z^\delta f(z) = \frac{1}{\Gamma(1-\delta)} \frac{d}{dz} \int_0^z \frac{f(t)}{(z-t)^\delta} dt, \quad 0 \le \delta < 1, \tag{2}$$

f is an analytic function in a simply connected region of the z-plane containing the origin and the multiplicity of $(z-t)^{-\delta}$ is removed by requiring $\log(z-t)$ to be real when $(z-t)$ is greater than zero. Clearly $f(z) = \lim_{\delta \to 0} D_z^\delta f(z)$ and

$$f'(z) = \lim_{\delta \to 1} D_z^\delta f(z).$$

* Email: shilpa2254@gmail.com
** Email: dileepl84@gmail.com
*** Email: drlatha@gmail.com

For the analytic function f of the form (1), Srivastava and Owa [2], introduced the operator Ω^δ and is defined by

$$\Omega^\delta f(z) = \Gamma(2-\delta)z^\delta D_z^\delta f(z)$$
$$= z + \sum_{n=2}^{\infty} k(n,\delta)a_n z^n$$

where $k(n,\delta) = \dfrac{n!\,\Gamma(2-\delta)}{\Gamma(n+1-\delta)}, \quad 0 \le \delta < 1.$

The fractional derivative of f can be expressed in terms of Hadamard product of two analytic functions in the open unit disc. Let f and $g \in \mathscr{A}$, where $f(z)$ is of the form (1) and $g(z) = z + \sum_{n=2}^{\infty} b_n z^n$, then $(f*g)(z) = z + \sum_{n=2}^{\infty} a_n b_n z^n$ is called the convolution or Hadamard product of f and g. Using the fractional derivative we define the following subclasses of $\mathscr{A}$.

Let $\Omega_\lambda^\delta(A,B)$ denote the class of analytic functions f obeying the condition

$$\frac{e^{i\lambda}\dfrac{z(\Omega^\delta f(z))'}{\Omega^\delta f(z)} - i\sin\lambda}{\cos\lambda} \in \mathscr{P}(A,B)$$

where $-1 \le B < A \le 1$, $0 \le \delta < 1$, $\lambda \ge 0$ and $z \in \mathscr{U}$.

For the parametric values $\delta = 0$ and $\delta = 1$ we get the classes $\mathscr{S}_\lambda^*(A,B)$ and $\mathscr{K}_\lambda(A,B)$ studied by Ganesan [3] respectively. For $\lambda = 0$ the class $\Omega_\lambda^\delta(A,B)$ reduces to the class $\Omega^\delta(A,B)$ consisting of analytic functions f such that

$$\frac{z(\Omega^\delta f(z))'}{\Omega^\delta f(z)} \in \mathscr{P}(A,B),$$

where $-1 \le B < A \le 1$, $0 \le \delta < 1$ and $z \in \mathscr{U}$.

For $\delta = 0$ and $\delta = 1$ we get the classes $\mathscr{S}^*(A,B)$ and $\mathscr{K}(A,B)$ studied by Ganesan[3] respectively.

2   Main Results

Theorem 1 *A function $f \in \mathscr{A}$ is of the form* (1), *is in the class* $\Omega^\delta(A,B)$ *if and only if*

$$\sum_{n=2}^{\infty} \frac{[n(1+B) - (1+A)]}{(B-A)} k(n,\delta)a_n < 1, \tag{3}$$

where $-1 \le B < A \le 1$, $0 \le \delta < 1$.

Proof. Suppose $f \in \Omega^\delta(A,B)$. Then

$$\frac{z(\Omega^\delta(A,B))'}{\Omega^\delta(A,B)} = \frac{1+A\omega(z)}{1+B\omega(z)}, \quad -1 \le B < A \le 1, \quad \delta \ge 0.$$

From this we get

$$\omega(z) = \frac{\Omega^\delta(A,B) - z(\Omega^\delta(A,B))'}{Bz(\Omega^\delta(A,B))' - A\Omega^\delta(A,B)} \quad \text{and} \quad |\omega(z)| < 1$$

implies

$$|\omega(z)| = \left| \frac{\displaystyle\sum_{n=2}^{\infty} k(n,\delta)a_n z^n (n-1)}{(B-A)z + \displaystyle\sum_{n=2}^{\infty} k(n,\delta)a_n z^n (Bn-A)} \right|. \tag{4}$$

Since $\Re\{\omega(z)\} \leq |\omega(z)| < 1.$

$$\Re\left\{ \frac{\displaystyle\sum_{n=2}^{\infty} k(n,\delta)a_n z^n (n-1)}{(B-A)z + \displaystyle\sum_{n=2}^{\infty} k(n,\delta)a_n z^n (Bn-A)} \right\} < 1.$$

Let z real, for sufficiently small z with $0 < z < 1$, the denominator of last expression is positive and so allowing $z \to 1$ in last inequality yields

$$\frac{\displaystyle\sum_{n=2}^{\infty} k(n,\delta)a_n (n-1)}{(B-A) + \displaystyle\sum_{n=2}^{\infty} k(n,\delta)a_n (Bn-A)} < 1,$$

$$\sum_{n=2}^{\infty} k(n,\delta)a_n (n-1) < (B-A) - \sum_{n=2}^{\infty} k(n,\delta)a_n (Bn-A).$$

On simplification we get (3).

Conversely, assume that

$$\sum_{n=2}^{\infty} [n(1+B) - (1+A)] k(n,\delta)a_n \leq (B-A).$$

We need to show that $|\omega(z)| < 1$, where

$$\frac{z(\Omega^\delta(A,B))'}{\Omega^\delta(A,B)} = \frac{1+A\omega(z)}{1+B\omega(z)}, \quad -1 \leq B < A \leq 1, \quad 0 \leq \delta < 1.$$

Since $|z| < 1$ by (4), we obtain

$$|\omega(z)| = \left| \frac{\sum\limits_{n=2}^{\infty} k(n,\delta)(n-1)a_n z^{n-1}}{(B-A) + \sum\limits_{n=2}^{\infty} k(n,\delta)(Bn-A)a_n z^{n-1}} \right|$$

$$\leq \frac{\sum\limits_{n=2}^{\infty} k(n,\delta)(n-1)a_n |z^{n-}|}{(B-A) + \sum\limits_{n=2}^{\infty} k(n,\delta)(Bn-A)a_n |z^{n-1}|}$$

$$< \frac{\sum\limits_{n=2}^{\infty} k(n,\delta)(n-1)a_n}{(B-A) + \sum\limits_{n=2}^{\infty} k(n,\delta)(Bn-A)a_n}.$$

The last expression is bounded by 1 if

$$\sum_{n=2}^{\infty} k(n,\delta)(n-1)a_n < (B-A) - \sum_{n=2}^{\infty} k(n,\delta)(Bn-A)a_n,$$

or

$$\sum_{n=2}^{\infty} [n(1+B) - (1+A)]\, k(n,\delta)a_n < (B-A),$$

which is true by hypothesis. Hence the proof.

Analogously we get the following Theorem for the class $\Omega_\lambda^\delta A, B$.

Theorem 2 *A function $f \in \mathscr{A}$ is of the form* (1), *is in the class $\Omega_\lambda^\delta(A,B)$ if and only if*

$$\sum_{n=2}^{\infty} \frac{[n(1+B) - (1+\gamma)]}{(B-\gamma)} k(n,\delta)a_n < 1, \tag{5}$$

where $\gamma = (A\cos\lambda + iB\sin\lambda)e^{-i\lambda}$, $-1 \leq B < A \leq 1$, $0 \leq \delta < 1$.

Theorem 3 *If f is of the form* (1) *and $g(z) = z + \sum\limits_{n=2}^{\infty} b_n z^n$ belongs to the class $\Omega_\lambda^\delta(A,B)$, then $h(z) = (f * g)(z) = z + \sum\limits_{n=2}^{\infty} a_n b_n z^n$ will be an element of $\Omega_\lambda^\delta(A_1,B_1)$ with $-1 \leq B_1 < A_1 \leq 1$, where $B_1 \geq \dfrac{\gamma_1 + \alpha}{1-\alpha}$, $\gamma_1 \leq 1 - 2\alpha$, where*

$$\alpha = \frac{(2-\delta)(B-\gamma)^2}{2[2B-\gamma+1]^2 - (2-\delta)(B-\gamma)^2}, \qquad \gamma = (A\cos\lambda + iB\sin\lambda)e^{-i\lambda}$$

and these bounds are sharp.

Proof. From (3), we have

$$\sum_{n=2}^{\infty} \frac{[n(1+B)-(1+\gamma)]}{(B-\gamma)} k(n,\delta)a_n < 1 \tag{6}$$

and

$$\sum_{n=2}^{\infty} \frac{[n(1+B)-(1+\gamma)]}{(B-\gamma)} k(n,\delta)b_n < 1. \tag{7}$$

In view of Cauchy-Schwarz inequality, from (6) and (7) we obtain

$$\sum_{n=2}^{\infty} uk(n,\delta)\sqrt{a_n b_n} \leq 1 \quad \text{where} \quad u = \frac{[n(1+B)-(1+\gamma)]}{(B-\gamma)}. \tag{8}$$

We need to find γ_1 and B_1 such that $h = f * g \in \Omega_{\lambda}^{\delta}(A_1, B_1)$, or equivalently,

$$\sum_{n=2}^{\infty} u_1 k(n,\delta)a_n b_n \leq 1 \quad \text{where} \quad u_1 = \frac{[n(1+B_1)-(1+\gamma_1)]}{(B_1-\gamma_1)}. \tag{9}$$

The inequality (7) will be true if $u_1 k(n,\delta)a_n b_n \leq uk(n,\delta)\sqrt{a_n b_n}$ or

$$\sqrt{a_n b_n} \leq \frac{u}{u_1}. \tag{10}$$

But from (6) we get $\sqrt{a_n b_n} \leq \dfrac{1}{uk(n,\delta)}$.

Thus (8) will be true if $\dfrac{1}{uk(n,\delta)} \leq \dfrac{u}{u_1}$, or $u_1 \leq u^2 k(n,\delta)$, that is,

$$\frac{[n(1+B_1)-(1+\gamma_1)]}{(B_1-\gamma_1)} \leq u^2 k(n,\delta). \tag{11}$$

Using $-1 \leq B < A \leq 1$, it is obvious that $u^2 k(n,\delta) > 1$, for $n \geq 2$ and hence (9) yields

$$\frac{(B_1-\gamma_1)}{1+B_1} \geq \frac{n-1}{u^2 k(n,\delta)-1} = \phi(n).$$

Since $\phi(n)$ is a decreasing function of n, for $n \geq 2$, $\phi(2)$ is the maximum value of $\phi(n)$ and therefore

$$\frac{B_1-\gamma_1}{1+B_1} \geq \frac{(2-\delta)(B-\gamma)^2}{2[2B-\gamma+1]^2-(2-\delta)(B-\gamma)^2} = \alpha. \tag{12}$$

Obviously, $\alpha < 1$ and fixing γ_1 in (10), we get

$$B_1 \geq \frac{\gamma_1+\alpha}{1-\alpha}. \tag{13}$$

Substituting $B_1 \leq 1$ in (11) we obtain, $\gamma_1 \leq 1 - 2\alpha$.

For verifying sharpness, we note that if $n = 2$, then

$$k(2,\delta) = \frac{2}{2-\delta}$$

and hence by taking

$$f(z) = g(z) = z + \frac{(B-\gamma)(2-\delta)}{2[2B-\gamma+1]}z^2 \in \Omega_\lambda^\delta(A,B)$$

we get

$$h(z) = (f*g)(z) = z + \frac{(B-\gamma)^2(2-\delta)^2}{4[2B-\gamma+1]^2}z^2.$$

If we use this function h in (7), the inequality (10) transforms to equality, that is, $\dfrac{B_1-\gamma_1}{1+B_1} = \alpha$. Now if $B_1 = 1$, then $\gamma_1 = 1 - \alpha$, which shows in this case $h \in \Omega_\lambda^\delta(1-2\alpha, 1)$.

As a special case of the above Theorem we get the following result.

Corollary 1 *If f is of the form (1) and $g(z) = z + \displaystyle\sum_{n=2}^{\infty} b_n z^n$ belongs to the class $\Omega^\delta(A,B)$, then $h(z) = (f*g)(z) = z + \displaystyle\sum_{n=2}^{\infty} a_n b_n z^n$ will be an element of $\Omega^\delta(A_1,B_1)$ with $-1 \leq B_1 < A_1 \leq 1$, where $B_1 \geq \dfrac{A_1+\alpha}{1-\alpha}$, $A \leq 1 - 2\alpha$, where*

$$\alpha = \frac{(2-\delta)(B-A)^2}{2[2B-A+1]^2 - (2-\delta)(B-A)^2}.$$

References

1. W. Janowski, Some extremal problems for certain families of analytic functions, I. Ann. Polon. Math., 28 (1973), 298–326.

2. H.M. Srivastava and S. Owa, An application of the fractional derivative, Math. Japan, 29 (1984), 383–389.

3. M.S. Ganesan and K.S. Padmanabhan, Convolution condition for certain classes of anlaytic functions, International J. Pure and Appl. Math., 15(7) (1984), 777–780.

Intuitionistic Fuzzy Translations of Intuitionistic Fuzzy H-ideals in BCK/BCI-algebras

Tapan Senapati[†], Monoranjan Bhowmik[‡,*] and Madhumangal Pal[†,**]

[†]Department of Applied Mathematics with Oceanology and Computer Programming,
Vidyasagar University, Midnapore – 721 102, West Bengal.
[‡]Department of Mathematics, V.T.T. College, Midnapore,
Paschim Medinipur – 721 101, West Bengal.

Abstract. Based on intuitionistic fuzzy set theory translations of H-ideals in BCK/BCI-algebras are discussed and several related properties are investigated.

Key words: Intuitionistic fuzzy ideal, intuitionistic fuzzy H-ideal, intuitionistic fuzzy translation.

1 Introduction

BCK-algebras and BCI-algebras are two important classes of logical algebras introduced by Imai and Iseki. It is known that the class of BCK-algebra is a proper subclass of the class of BCI-algebras. The concept of fuzzy translations in fuzzy subalgebras and ideals in BCK/BCI-algebras has been discussed respectively by Lee et al. [1] and Jun [2]. They investigated relations among fuzzy translations, fuzzy extensions and fuzzy multiplications. Motivated by this, in [3], the authors have studied fuzzy translations of fuzzy H-ideals in BCK/BCI-algebras. They also extend this study from fuzzy translations to intuitionistic fuzzy translations [4] in BCK/BCI-algebras. In this paper intuitionistic fuzzy translations of intuitionistic fuzzy H-ideals in BCK/BCI-algebras are discussed.

2 Preliminaries

By a BCI-algebra we mean an algebra $(X, *, 0)$ of type $(2, 0)$ satisfying the following axioms for all $x, y, z \in X$:

    (i) $((x*y)*(x*z))*(z*y) = 0$
    (ii) $(x*(x*y))*y = 0$
    (iii) $x*x = 0$
    (iv) $x*y = 0$ and $y*x = 0$ imply $x = y$.

If a BCI-algebra X satisfies $0*x = 0$ for all $x \in X$, then we say that X is a BCK-algebra. We can define a partial ordering "$\leq$" by $x \leq y$ if and only if

  * Email: mbvttc@gmail.com
  ** Email: mmpalvu@gmail.com

$x * y = 0$. Throughout this paper, X always means a BCK/BCI-algebra without any specification.

A non-empty subset S of X is called a subalgebra of X if $x * y \in S$ for any $x, y \in S$. A nonempty subset I of X is called an ideal of X if it satisfies (I_1) $0 \in I$ and (I_2) $x * y \in I$ and $y \in I$ imply $x \in I$. A non-empty subset I of X is said to be a H-ideal [5] of X if it satisfies (I_1) and (I_3) $x * (y * z) \in I$ and $y \in I$ imply $x * z \in I$ for all $x, y, z \in X$. A BCI-algebra is said to be associative if $(x * y) * z = x * (y * z)$ for all $x, y, z \in X$.

An intuitionistic fuzzy set $A = \{\langle x, \mu_A(x), \nu_A(x)\rangle : x \in X\}$ in X is called an intuitionistic fuzzy subalgebra of X if it satisfies the two conditions (F_1) $\mu_A(x * y) \geq \min\{\mu_A(x), \mu_A(y)\}$ and (F_2) $\nu_A(x * y) \leq \max\{\nu_A(x), \nu_A(y)\}$ for all $x, y \in X$.

An intuitionistic fuzzy set $A = \{\langle x, \mu_A(x), \nu_A(x)\rangle : x \in X\}$ in X is called an intuitionistic fuzzy ideal of X if it satisfies (F_3) $\mu_A(0) \geq \mu_A(x)$, $\nu_A(0) \leq \nu_A(x)$, (F_4) $\mu_A(x) \geq \min\{\mu_A(x * y), \mu_A(y)\}$ and (F_5) $\nu_A(x) \leq \max\{\nu_A(x * y), \nu_A(y)\}$ for all $x, y \in X$.

An intuitionistic fuzzy set A in X is called an intuitionistic fuzzy H-ideal [6] of X if it satisfies (F_6) and (F_6) $\mu_A(x * z) \geq \min\{\mu_A(x * (y * z)), \mu_A(y)\}$ (F_7) $\nu_A(x * z) \leq \max\{\nu_A(x * (y * z)), \nu_A(y)\}$ for all $x, y, z \in X$.

3   Main Results

For the sake of simplicity, we shall use the symbol $A = (\mu_A, \nu_A)$ for the intuitionistic fuzzy subset $A = \{\langle x, \mu_A(x), \nu_A(x)\rangle : x \in X\}$. Throughout this paper, we take $\mathfrak{T} := \inf\{\nu_A(x) | x \in X\}$ for any intuitionistic fuzzy set $A = (\mu_A, \nu_A)$ of X.

Definition 1. [4] Let $A = (\mu_A, \nu_A)$ be an intuitionistic fuzzy subset of X and let $\alpha \in [0, \mathfrak{T}]$. An object having the form $A_\alpha^T = ((\mu_A)_\alpha^T, (\nu_A)_\alpha^T)$ is called an intuitionistic fuzzy α-translation of A if $(\mu_A)_\alpha^T(x) = \mu_A(x) + \alpha$ and $(\nu_A)_\alpha^T(x) = \nu_A(x) - \alpha$ for all $x \in X$.

Theorem 1. *If $A = (\mu_A, \nu_A)$ is an intuitionistic fuzzy H-ideals of X, then the intuitionistic fuzzy α-translation $A_\alpha^T = ((\mu_A)_\alpha^T, (\nu_A)_\alpha^T)$ of A is an intuitionistic fuzzy H-ideals of X for all $\alpha \in [0, \mathfrak{T}]$.*

Proof. Let $A = (\mu_A, \nu_A)$ is an intuitionistic fuzzy H-ideals of X and $\alpha \in [0, \mathfrak{T}]$. Then $(\mu_A)_\alpha^T(0) = \mu_A(0) + \alpha \geq \mu_A(x) + \alpha = (\mu_A)_\alpha^T(x)$ and $(\nu_A)_\alpha^T(0) = \nu_A(0) - \alpha \leq \nu_A(x) - \alpha = (\nu_A)_\alpha^T(x)$ for all $x \in X$. Now, $(\mu_A)_\alpha^T(x * z) = \mu_A(x * z) + \alpha \geq \min\{\mu_A(x * (y * z)), \mu_A(y)\} + \alpha = \min\{\mu_A(x * (y * z)) + \alpha, \mu_A(y) + \alpha\} = \min\{(\mu_A)_\alpha^T(x * (y * z)), (\mu_A)_\alpha^T(y)\}$ and similarly $(\nu_A)_\alpha^T(x * z) \leq \max\{(\nu_A)_\alpha^T(x * (y * z)), (\nu_A)_\alpha^T(y)\}$ for all $x, y, z \in X$. Hence A_α^T of A is an intuitionistic fuzzy H-ideals of X.

Theorem 2. *Let $A = (\mu_A, \nu_A)$ be an intuitionistic fuzzy subset of X such that the intuitionistic fuzzy α-translation $A_\alpha^T = ((\mu_A)_\alpha^T, (\nu_A)_\alpha^T)$ of A is an intuitionistic fuzzy H-ideals of X for some $\alpha \in [0, \mathfrak{T}]$. Then $A = (\mu_A, \nu_A)$ is an intuitionistic fuzzy H-ideal of X.*

Theorem 3. *If the intuitionistic fuzzy α-translation $A_\alpha^T = ((\mu_A)_\alpha^T, (\nu_A)_\alpha^T)$ of A is an intuitionistic fuzzy H-ideal of X for all $\alpha \in [0, \mathfrak{T}]$ then it must be an intuitionistic fuzzy subalgebra of X.*

Proof. Let the intuitionistic fuzzy α-translation $A_\alpha^T = ((\mu_A)_\alpha^T, (\nu_A)_\alpha^T)$ of A is an intuitionistic fuzzy H-ideal of X. Then we have $(\mu_A)_\alpha^T(x*z) \geq \min\{(\mu_A)_\alpha^T(x*(y*z)), (\mu_A)_\alpha^T(y)\}$ and $(\nu_A)_\alpha^T(x*z) \leq \max\{(\nu_A)_\alpha^T(x*(y*z)), (\nu_A)_\alpha^T(y)\}$ for all $x, y, z \in X$. Substituting y for z we get $(\mu_A)_\alpha^T(x*y) \geq \min\{(\mu_A)_\alpha^T(x*(y*y)), (\mu_A)_\alpha^T(y)\} = \min\{(\mu_A)_\alpha^T(x*0), (\mu_A)_\alpha^T(y)\} = \min\{(\mu_A)_\alpha^T(x), (\mu_A)_\alpha^T(y)\}$ and similarly $(\nu_A)_\alpha^T(x*y) \leq \max\{(\nu_A)_\alpha^T(x), (\nu_A)_\alpha^T(y)\}$. Therefore, A_α^T is an intuitionistic fuzzy subalgebra of X.

Proposition 1. *Let $A = (\mu_A, \nu_A)$ be an intuitionistic fuzzy subset of X such that the intuitionistic fuzzy α-translation $A_\alpha^T = ((\mu_A)_\alpha^T, (\nu_A)_\alpha^T)$ of A is an intuitionistic fuzzy ideal of X for $\alpha \in [0, \mathfrak{T}]$. If $(x*a)*b = 0$ for all $a, b, x \in X$, then $(\mu_A)_\alpha^T(x) \geq \min\{(\mu_A)_\alpha^T(a), (\mu_A)_\alpha^T(b)\}$ and $(\nu_A)_\alpha^T(x) \leq \max\{(\nu_A)_\alpha^T(a), (\nu_A)_\alpha^T(b)\}$.*

Theorem 4. *Let $A = (\mu_A, \nu_A)$ be an intuitionistic fuzzy subset of X such that the intuitionistic fuzzy α-translation $A_\alpha^T = ((\mu_A)_\alpha^T, (\nu_A)_\alpha^T)$ of A is an intuitionistic fuzzy ideal of X for $\alpha \in [0, \mathfrak{T}]$. If it satisfies the condition $(\mu_A)_\alpha^T(x*y) \geq (\mu_A)_\alpha^T(x)$ and $(\nu_A)_\alpha^T(x*y) \leq (\nu_A)_\alpha^T(x)$ for all $x, y \in X$, then A_α^T of A is an intuitionistic fuzzy H-ideal of X.*

Proof. Let the intuitionistic fuzzy α-translation A_α^T of A is an intuitionistic fuzzy ideal of X. For any $x, y, z \in X$, we have $(\mu_A)_\alpha^T(x*z) \geq \min\{(\mu_A)_\alpha^T((x*z)*(y*z)), (\mu_A)_\alpha^T(y*z)\} = \min\{(\mu_A)_\alpha^T((x*(y*z))*z), (\mu_A)_\alpha^T(y*z)\} \geq \min\{(\mu_A)_\alpha^T(x*(y*z)), (\mu_A)_\alpha^T(y)\}$ and similarly $(\nu_A)_\alpha^T(x*z) \leq \max\{(\nu_A)_\alpha^T(x*(y*z)), (\nu_A)_\alpha^T(y)\}$. Hence, A_α^T of A is an intuitionistic fuzzy H-ideal of X for some $\alpha \in [0, \mathfrak{T}]$.

Theorem 5. *If $A = (\mu_A, \nu_A)$ be an intuitionistic fuzzy subset of associative BCK/BCI-algebra X such that the intuitionistic fuzzy α-translation $A_\alpha^T = ((\mu_A)_\alpha^T, (\nu_A)_\alpha^T)$ of A is an intuitionistic fuzzy ideal of X for $\alpha \in [0, \mathfrak{T}]$, then A_α^T of A is an intuitionistic fuzzy H-ideal of X.*

Proof. Let the intuitionistic fuzzy α-translation A_α^T of A is an intuitionistic fuzzy ideal of X. For any $x, y, z \in X$, we have $(\mu_A)_\alpha^T(x*z) \geq \min\{(\mu_A)_\alpha^T((x*z)*y), (\mu_A)_\alpha^T(y)\} = \min\{(\mu_A)_\alpha^T((x*y)*z), (\mu_A)_\alpha^T(y)\} = \min\{(\mu_A)_\alpha^T(x*(y*z)), (\mu_A)_\alpha^T(y)\}$ and $(\nu_A)_\alpha^T(x*z) \leq \max\{(\nu_A)_\alpha^T((x*z)*y), (\nu_A)_\alpha^T(y)\} = \max\{(\nu_A)_\alpha^T((x*y)*z), (\nu_A)_\alpha^T(y)\} = \max\{(\nu_A)_\alpha^T(x*(y*z)), (\nu_A)_\alpha^T(y)\}$. Hence, A_α^T is an intuitionistic fuzzy H-ideal of X.

Theorem 6. *If $A = (\mu_A, \nu_A)$ be an intuitionistic fuzzy subset of X such that the intuitionistic fuzzy α-translation $A_\alpha^T = ((\mu_A)_\alpha^T, (\nu_A)_\alpha^T)$ of A is an intuitionistic fuzzy H-ideal of X for $\alpha \in [0, \mathfrak{T}]$, then the sets $T_{\mu_A} := \{x \in X | (\mu_A)_\alpha^T(x) = (\mu_A)_\alpha^T(0)\}$ and $T_{\nu_A} := \{x \in X | (\nu_A)_\alpha^T(x) = (\nu_A)_\alpha^T(0)\}$ are H-ideals of X.*

Proof. Suppose that $A_\alpha^T = ((\mu_A)_\alpha^T, (\nu_A)_\alpha^T)$ is an intuitionistic fuzzy ideal of X. Then $(\mu_A)_\alpha^T$ and $(\nu_A)_\alpha^T$ are fuzzy H-ideal of X. Obviously $0 \in T_{\mu_A}, T_{\nu_A}$. Let $x, y, z \in X$ be such that $x * (y * z) \in T_{\mu_A}$ and $y \in T_{\mu_A}$. Then $(\mu_A)_\alpha^T(x * (y * z)) = (\mu_A)_\alpha^T(0) = (\mu_A)_\alpha^T(y)$ and so $(\mu_A)_\alpha^T(x * z) \geq \min\{(\mu_A)_\alpha^T(x * (y * z)), (\mu_A)_\alpha^T(y)\} = (\mu_A)_\alpha^T(0)$. Since, $(\mu_A)_\alpha^T$ is a fuzzy H-ideal of X, we conclude that $(\mu_A)_\alpha^T(x * z) = (\mu_A)_\alpha^T(0)$. This implies $\mu_A(x * z) + \alpha = \mu_A(0) + \alpha$ or, $\mu_A(x * z) = \mu_A(0)$ so that $x * z \in T_{\mu_A}$. Therefore T_{μ_A} is a H-ideal of X.

Again let $a, b, c \in X$ be such that $a * (b * c) \in T_{\nu_A}$ and $b \in T_{\nu_A}$. Then $(\nu_A)_\alpha^T(a * (b * c)) = (\nu_A)_\alpha^T(0) = (\nu_A)_\alpha^T(b)$ and so $(\nu_A)_\alpha^T(a * c) \leq \max\{(\nu_A)_\alpha^T(a * (b * c)), (\nu_A)_\alpha^T(b)\} = (\nu_A)_\alpha^T(0)$. Since, $(\nu_A)_\alpha^T$ is a fuzzy H-ideal of X, we conclude that $(\nu_A)_\alpha^T(a * c) = (\nu_A)_\alpha^T(0)$. This implies $\nu_A(a * c) + \alpha = \nu_A(0) + \alpha$ or, $\nu_A(a * c) = \nu_A(0)$ so that $a * c \in T_{\nu_A}$. Therefore T_{ν_A} is a H-ideal of X.

Proposition 2. *Let the intuitionistic fuzzy α-translation $A_\alpha^T = ((\mu_A)_\alpha^T, (\nu_A)_\alpha^T)$ of A be an intuitionistic fuzzy H-ideal of X for $\alpha \in [0, \mathfrak{T}]$. If $x \leq y$ then $(\mu_A)_\alpha^T(x) \geq (\mu_A)_\alpha^T(y)$ and $(\nu_A)_\alpha^T(x) \leq (\nu_A)_\alpha^T(y)$, that is, $(\mu_A)_\alpha^T$ is order reversing and $(\nu_A)_\alpha^T$ is order prereversing.*

Theorem 7. *Let $A = (\mu_A, \nu_A)$ be an intuitionistic fuzzy subset of X such that the intuitionistic fuzzy α-translation $A_\alpha^T = ((\mu_A)_\alpha^T, (\nu_A)_\alpha^T)$ of A is an intuitionistic fuzzy ideal of X for $\alpha \in [0, \mathfrak{T}]$, then the following assertions are equivalent:*
(i) A_α^T is an intuitionistic fuzzy H-ideal of X,
(ii) $(\mu_A)_\alpha^T(x * y) \geq (\mu_A)_\alpha^T(x * (0 * y))$ and $(\nu_A)_\alpha^T(x * y) \leq (\nu_A)_\alpha^T(x * (0 * y))$ for all $x, y \in X$,
(iii) $(\mu_A)_\alpha^T((x * y) * z) \geq (\mu_A)_\alpha^T(x * (y * z))$ and $(\nu_A)_\alpha^T((x * y) * z) \leq (\nu_A)_\alpha^T(x * (y * z))$ for all $x, y, z \in X$.

Proof. Assume that $A_\alpha^T = ((\mu_A)_\alpha^T, (\nu_A)_\alpha^T)$ is an intuitionistic fuzzy ideal of X. Then $(\mu_A)_\alpha^T$ and $(\nu_A)_\alpha^T$ are fuzzy H-ideal of X.

$(i) \Rightarrow (ii)$ Let A_α^T is an intuitionistic fuzzy H-ideal of X. Then for all $x, y \in X$ we have $(\mu_A)_\alpha^T(x * y) \geq \min\{(\mu_A)_\alpha^T(x * (0 * y)), (\mu_A)_\alpha^T(0)\} = (\mu_A)_\alpha^T(x * (0 * y))$ and $(\nu_A)_\alpha^T(x * y) \leq \max\{(\nu_A)_\alpha^T(x * (0 * y)), (\nu_A)_\alpha^T(0)\} = (\nu_A)_\alpha^T(x * (0 * y))$. Therefore the inequality (ii) is satisfied.

$(ii) \Rightarrow (iii)$ Assume that (ii) is satisfied. For all $x, y, z \in X$, we have $((x * y) * (0 * z)) * (x * (y * z)) = ((x * y) * (x * (y * z))) * (0 * z) \leq ((y * z) * y) * (0 * z) = ((y * y) * z) * (0 * z) = (0 * z) * (0 * z) = 0$. It follows from Proposition 2 that $(\mu_A)_\alpha^T((x * y) * (0 * z)) * (x * (y * z)) \geq (\mu_A)_\alpha^T(0)$ and $(\nu_A)_\alpha^T((x * y) * (0 * z)) * (x * (y * z)) \leq (\nu_A)_\alpha^T(0)$. Since $(\mu_A)_\alpha^T$ and $(\nu_A)_\alpha^T$ are fuzzy H-ideal of X, therefore, we have $(\mu_A)_\alpha^T((x * y) * (0 * z)) * (x * (y * z)) = (\mu_A)_\alpha^T(0)$ and $(\nu_A)_\alpha^T((x * y) * (0 *$

$z)) * (x * (y * z)) = (v_A)_\alpha^T(0)$. Using (ii) we get $(\mu_A)_\alpha^T((x * y) * z) \geq (\mu_A)_\alpha^T((x * y) * (0 * z)) = \min\{(\mu_A)_\alpha^T(((x * y) * (0 * z)) * (x * (y * z))), (\mu_A)_\alpha^T(x * (y * z))\} = \min\{(\mu_A)_\alpha^T(0), (\mu_A)_\alpha^T(x * (y * z))\} = (\mu_A)_\alpha^T(x * (y * z))$ and similarly $(v_A)_\alpha^T((x * y) * z) \leq (v_A)_\alpha^T(x * (y * z))$. Therefore inequality (iii) is also satisfied.

$(iii) \Rightarrow (i)$ Assume that (iii) is valid. For all $x, y, z \in X$, we have $(\mu_A)_\alpha^T(x * z) \geq \min\{(\mu_A)_\alpha^T((x * z) * y), (\mu_A)_\alpha^T(y)\} = \min\{(\mu_A)_\alpha^T((x * y) * z), (\mu_A)_\alpha^T(y)\} \geq \min\{(\mu_A)_\alpha^T(x * (y * z)), (\mu_A)_\alpha^T(y)\}$ and similarly $(v_A)_\alpha^T(x * z) \leq \max\{(v_A)_\alpha^T(x * (y * z)), (v_A)_\alpha^T(y)\}$. Therefore A_α^T is an intuitionistic fuzzy H-ideal of X. Hence, the assertion (i) holds.

Theorem 8. *Let $A = (\mu_A, v_A)$ be an intuitionistic fuzzy subset of X such that the intuitionistic fuzzy α-translation $A_\alpha^T = ((\mu_A)_\alpha^T, (v_A)_\alpha^T)$ of A is an intuitionistic fuzzy ideal of X, then the following assertions are equivalent:*
(i) A_α^T is an intuitionistic fuzzy H-ideal of X,
(ii) $(\mu_A)_\alpha^T((x * z) * y) \geq (\mu_A)_\alpha^T((x * z) * (0 * y))$ and $(v_A)_\alpha^T((x * z) * y) \leq (v_A)_\alpha^T((x * z) * (0 * y))$ for all $x, y, z \in X$,
(iii) $(\mu_A)_\alpha^T(x * y) \geq \min\{(\mu_A)_\alpha^T((x * z) * (0 * y)), (\mu_A)_\alpha^T(z)\}$ and $(v_A)_\alpha^T(x * y) \leq \max\{(v_A)_\alpha^T((x * z) * (0 * y)), (v_A)_\alpha^T(z)\}$ for all $x, y, z \in X$.

Proof. $(i) \Rightarrow (ii)$ Same as above theorem.

$(ii) \Rightarrow (iii)$ Assume that (ii) is valid. For all $x, y, z \in X$, we have $(\mu_A)_\alpha^T(x * y) \geq \min\{(\mu_A)_\alpha^T((x * y) * z), (\mu_A)_\alpha^T(z)\} = \min\{(\mu_A)_\alpha^T((x * z) * y), (\mu_A)_\alpha^T(z)\} \geq \min\{(\mu_A)_\alpha^T((x * z) * (0 * y)), (\mu_A)_\alpha^T(z)\}$ and $(v_A)_\alpha^T(x * y) \leq \max\{(v_A)_\alpha^T((x * y) * z), (v_A)_\alpha^T(z)\} = \max\{(v_A)_\alpha^T((x * z) * y), (v_A)_\alpha^T(z)\} \leq \max\{(v_A)_\alpha^T((x * z) * (0 * y)), (v_A)_\alpha^T(z)\}$.

Therefore (iii) is satisfied.

$(iii) \Rightarrow (i)$ Assume that (iii) is valid. Therefore for all $x, y, z \in X$, we have $(\mu_A)_\alpha^T(x * y) \geq \min\{(\mu_A)_\alpha^T((x * z) * (0 * y)), (\mu_A)_\alpha^T(z)\}$ and $(v_A)_\alpha^T(x * y) \leq \max\{(v_A)_\alpha^T((x * z) * (0 * y)), (v_A)_\alpha^T(z)\}$. Putting $z = 0$ we get $(\mu_A)_\alpha^T(x * y) \geq \min\{(\mu_A)_\alpha^T((x * 0) * (0 * y)), (\mu_A)_\alpha^T(0)\} = \min\{(\mu_A)_\alpha^T(x * (0 * y)), (\mu_A)_\alpha^T(0)\} = (\mu_A)_\alpha^T(x * (0 * y))$ and $(v_A)_\alpha^T(x * y) \leq \max\{(v_A)_\alpha^T((x * 0) * (0 * y)), (v_A)_\alpha^T(0)\} = \max\{(v_A)_\alpha^T(x * (0 * y)), (v_A)_\alpha^T(0)\} = (v_A)_\alpha^T(x * (0 * y))$. It follows from Theorem 7 that A_α^T is an intuitionistic fuzzy H-ideal of X.

References

1. K.J. Lee, Y.B. Jun and M.I. Doh, Fuzzy translations and fuzzy multiplications of *BCK/BCI*-algebras, Commun. Korean Math. Soc., 24(3)(2009), 353–360.
2. Y.B. Jun, Translations of fuzzy ideals in *BCK/BCI*-algebras, Hacettepe J. of Mathematics and Statistics, 40(3)(2011), 349–358.
3. T. Senapati, M. Bhowmik, M. Pal and B. Davvaz, Fuzzy translations of fuzzy *H*-ideals in *BCK/BCI*-algebras, Communicated.
4. T. Senapati, M. Bhowmik, M. Pal and B. Davvaz, Atanassov's intuitionistic fuzzy translations of intuitionistic fuzzy subalgebras and ideals in *BCK/BCI*-algebras, Communicated.

5. H.M. Khalid and B. Ahmad, Fuzzy H-ideals in BCI-algebras, Fuzzy Sets and Systems, 101(1999), 153–158.

6. B. Satyanarayana, U.B. Madhavi and R.D. Prasad, On intuitionistic fuzzy H-Ideals in BCK-algebras, International Journal of Algebra, 4(15)(2010), 743–749.

Generalized Intuitionistic Fuzzy Nilpotent Matrices over Distributive Lattice

Amal Kumar Adak[†], Monoranjan Bhowmik[‡,*] and Madhumangal Pal[†,**]

[†]Department of Applied Mathematics with Oceanology and Computer Programming,
Vidyasagar University, Midnapore – 721 102, West Bengal.
[‡]Department of Mathematics, V.T.T. College, Midnapore,
Paschim Medinipur – 721 101, West Bengal.

Abstract. In this paper, the concept of lattice over generalized intuitionistic fuzzy matrices (GIFMs) are introduced and have shown that the set of GIFMs forms a distributive lattice. Some algebric properties of generalized intuitionistic fuzzy matrices (GIFMs) over distributive lattice are presented. Also, some characteristics of generalized intuitionistic fuzzy nilpotent matrices (IFNMs) over distributive lattice are discussed.

Key words: Intuitionistic fuzzy matrices, generalized intuitionistic fuzzy matrices, distributive lattice of generalized intuitionistic fuzzy matrix, generalized intuitionistic fuzzy nilpotent matrices.

1   Introduction

Atanassov [1] introduced the concept of intuitionistic fuzzy sets (IFSs). Also a lot of reseach works done by several researchers on the field of IFS. Ragab and Emam [2] defined adjoint of a square fuzzy matrix. By the concept of IFSs, first time Pal [3] introduced intuitionistic fuzzy determinant. Latter on Pal and Shyamal [4] introduced intuitionistic fuzzy matrices and determine distance between intuitionistic fuzzy matrices. Mondal and Samanta [5] introduced an another concept of IFSs called generalized IFSs. Bhowmik and Pal [6] defined generalized interval-valued intuitionistic fuzzy set (GIVIFS) and presented various properties of it.

Lattice matrices are useful tools in various domains like the theory of switching, automata theory and theory of finite graphs. The notions of nilpotent lattice matrices seem to be appeared first in the work of Give'on [7]. In [7], Give'on proved that an $n \times n$ lattice matrix is nilpotent if and only if $A^n = \mathbf{0}$. Since then, a number of researchers have studied the topic of the nilpotent lattice matrices.

Our aim is to introduce and study distributive lattice over GIFMs. The structure of this paper is organized as follows. In Section 2, the preliminaries and some definitions are given. In Section 3, some algebric structures of GIFMs

---

* Email: mbvttc@gmail.com
** Email: mmpalvu@gmail.com

over distributive lattice are supplied and some results are given. In Section 4, some properties of generalized intuitionistic fuzzy determinant over distributive lattice (GIFD) are persented. In Section 5, the definition of generalized intuitionistic fuzzy nilpotent matrix (GIFNM) over distributive lattice is given.

2 Preliminaries

Here some preliminaries, definitions of IFSs and GIFMs are recalled and presented some algebric operations of GIFMs and different types of GIFMs.

2.1 Intuitionistic fuzzy set

Definition 1. *(Intuitionistic fuzzy set (IFS))* *An instuitionistic fuzzy set (IFS)* *A over X is an object having the form* $A = \left\{ x, \langle \mu_A(x), v_A(x) \rangle : x \in X \right\}$; *where* $\mu_A : X \to [0,1]$ *and* $v_A : X \to [0,1]$, *where* $\mu_A(x)$ *and* $v_A(x)$ *are called the membership and non-membership values of x in A satisfying the condition* $0 \leq \mu_A(x) + v_A(x) \leq 1$.

2.2 Intuitionistic fuzzy matrix(IFM) and Generalized intuitionistic fuzzy matrix (GIFM)

Definition 2. *(Intuitionistic fuzzy matrix (IFM))* An intuitionistic fuzzy matrix (IFM) of order $m \times n$ is defined as $A = \left[\langle a_{ij\mu}, a_{ijv} \rangle \right]$ where $a_{ij\mu}$ and a_{ijv} are the membership and non-membership values of the ij-th element in A satisfying the condition $0 \leq a_{ij\mu} + a_{ijv} \leq 1$ for all i, j.

Definition 3. *(Generalized intuitionistic fuzzy matrix (GIFM))* A generalised intuitionistic fuzzy matrix (GIFM) of order $m \times n$ is defined as $A = \left[\langle a_{ij\mu}, a_{ijv} \rangle \right]$ where $a_{ij\mu}$ and a_{ijv} are the membership and non-membership values of the ij-th element in A satisfying the generalized intuitionistic fuzzy condition $0 \leq a_{ij\mu} \wedge a_{ijv} \leq 0.5$ for all i, j.

Let $G_{m \times n}$ denotes the set of all GIFMs of order $m \times n$. In particular G_n denotes the set of all GIFMs of order $n \times n$.

Definition 4. *(Comparable GIFMs) Let A and B be two GIFMs such that* $A = \left[\langle a_{ij\mu}, a_{ijv} \rangle \right]$ *and* $B = \left[\langle b_{ij\mu}, b_{ijv} \rangle \right] \in G_{m \times n}$, *then two matrices A and B are said to be comparable GIFMs if* $a_{ij\mu} \leq b_{ij\mu}$ *and* $a_{ijv} \geq b_{ijv}$ *for all* i, j.

Some algebric operations of GIFMs

Let A and B be two GIFMs, such that $A = \left[\langle a_{ij\mu}, a_{ijv}\rangle\right]$ and $B = \left[\langle b_{ij\mu}, b_{ijv}\rangle\right] \in G_{m\times n}$,

1. Matrix addition and subtraction are given by

$$A + B = \left[\langle \max\{a_{ij\mu}, b_{ij\mu}\}, \min\{a_{ijv}, b_{ijv}\}\rangle\right]$$

$$\text{and } A - B = \left[\langle a_{ij\mu} - b_{ij\mu}, a_{ijv} - b_{ijv}\rangle\right],$$

where $a_{ij\mu} - b_{ij\mu} = \begin{cases} a_{ij\mu}, & a_{ij\mu} \geq b_{ij\mu} \\ 0, & \text{elsewhere,} \end{cases}$ and $a_{ijv} - b_{ijv} = \begin{cases} a_{ijv}, & a_{ijv} < b_{ijv} \\ 0, & \text{otherwise.} \end{cases}$

2. Componentwise matrix multiplication is given by
$$A \odot B = \left[\langle \min\{a_{ij\mu}, b_{ij\mu}\}, \max\{a_{ijv}, b_{ijv}\}\rangle\right].$$

3. Let A, B be two GIFMs of order $m \times n$ and $n \times p$. Then the matrix product AB is given by $AB = \left[\langle \sum_k \min\{a_{ik\mu}, b_{kj\mu}\}, \prod_k \max\{a_{ikv}, b_{kjv}\}\rangle\right] \in G_{m\times p}$.

2.3   Poset of GIFMs

Lemma 1. *(Poset of GIFMs) Let G_n be the set of all $n \times n$ GIFMs and '$\leq$' be comparable fuzzy matrix relation, then $(G_n, \leq)$ is a poset.*
(1) $A \leq A$ is true since $a_{ij\mu} \leq a_{ij\mu}$ and $a_{ijv} \geq a_{ijv}$. Hence the relation '$\leq$' is reflexive.
(2) $A \leq B$ and $B \leq A$ possible only when $A = B$, since $A \leq B$ when $a_{ij\mu} \leq b_{ij\mu}$ and $a_{ijv} \geq b_{ijv}$ and $B \leq A$ when $b_{ij\mu} \leq a_{ij\mu}$ and $b_{ijv} \geq a_{ijv}$. Combining above results give $A = B$. Therefore the relation '$\leq$' is anti-symmetric.
(3) $A \leq B$ and $B \leq C$ when $a_{ij\mu} \leq b_{ij\mu}$ and $a_{ijv} \geq b_{ijv}$ and $B \leq C$ when $b_{ij\mu} \leq a_{ij\mu}$ and $b_{ijv} \geq a_{ijv}$. Then it is obvious that $A \leq C$ since $a_{ij\mu} \leq c_{ij\mu}$ and $a_{ijv} \geq c_{ijv}$. Hence the relation '$\leq$' is transitive.

3   Distributive Lattice of GIFMs

In this section we introduce the concept of distributive lattice of GIMFs and give some properties of GIFNMs over distributive lattice. We begin this section with some definitions:

3.1   Lattice of GIFMs

A non-empty poset $(G_n, \leq)$ with two binary operation $+$ and $\odot$ is called a lattice if the following axioms hold:
(1) Closure : $A, B \in G_n$ then $A + B \in G_n$ and $A \odot B \in F_n$.
(2) Commutative : $A, B \in G_n$ then $A + B = B + A$ and $A \odot B = B \odot A$.

(3) Associative : $A, B, C \in G_n$ then $(A+B)+C = A+(B+C)$ and $(A \odot B) \odot C = A \odot (B \odot C)$.

(4) Absorption : $A, B \in G_n$ then $A \odot (A+B) = A$ and $A + (A \odot B) = A$.

Therefore, the poset $(G_n, \leq)$ with two binary operation matrix addition and componentwise matrix multiplication of GIFMs form lattice.

It should be noted that the poset $(G_n, \leq)$ with two binary operation matrix addition and matrix product of GIFMs does not form lattice as matrix product is not commutative.

Idempotent law: Let A be an $n \times n$ GIFMs over distributive lattice $(G_n(L), \leq, +, \odot)$, then A satisfy idempotent law i.e., (i) $A + A = A$ and (ii) $A \odot A = A$.

Theorem 1. *Let A, B be two square GIFMs of $n \times n$ over distributive lattice $(G_n(L), \leq, +, \odot)$ then $A \odot B = A$ if and only if $A + B = B$.*

Proof. Let $A \odot B = A$ where $A, B \in G_n(L)$

Therefore, $\min\{a_{ij\mu}, b_{ij\mu}\} = a_{ij\mu}$ and $\max\{a_{ij\nu}, b_{ij\nu}\} = a_{ij\nu}$.

Hence, $\max\{a_{ij\mu}, b_{ij\mu}\} = b_{ij\mu}$ and $\min\{a_{ij\nu}, b_{ij\nu}\} = b_{ij\nu}$.

Now, $A + B = \left[\left\langle \max\{a_{ij\mu}, b_{ij\mu}\}, \min\{a_{ij\nu}, b_{ij\nu}\} \right\rangle \right] = \left[\left\langle b_{ij\mu}, b_{ij\nu} \right\rangle \right] = B$.

Similarly, it can be proved the converse part of the theorem.

Theorem 2. *Let $(G_n(L), \leq, +, \odot)$ be the lattice of GIFMs and A, B, $C \in G_n$. If $A \leq B$ and $A \leq C$ then (1) $A \leq B+C$, (2) $A \leq B \odot C$.*

Proof. If $A \leq B$ then we have $a_{ij\mu} \leq b_{ij\mu}$ and $a_{ij\nu} \geq b_{ij\nu}$.

Again, $A \leq C$ we have $a_{ij\mu} \leq c_{ij\mu}$ and $a_{ij\mu} \geq c_{ij\nu}$.

Hence, $a_{ij\mu} \leq \max\{b_{ij\mu}, c_{ij\mu}\}$ and $a_{ij\nu} \geq \min\{b_{ij\nu}, c_{ij\nu}\}$.

Therefore, $A = \left[\left\langle a_{ij\mu}, a_{ij\nu} \right\rangle \right] \leq \left[\left\langle \max\{b_{ij\mu}, c_{ij\mu}\}, \min\{b_{ij\nu}, c_{ij\nu}\} \right\rangle \right] = B+C$

Similarly, it can be proved the second part of the theorem.

Theorem 3. *Let $(G_n(L), \leq, +, \odot)$ be a lattice over GIFMs and A, B, C, $D \in G_n$. If $A \leq B$ and $C \leq D$ then (1) $A+C \leq B+D$ and (2) $A \odot C \leq B \odot D$.*

Proof. If $A \leq B$ then we have $a_{ij\mu} \leq b_{ij\mu}$ and $a_{ij\nu} \geq b_{ij\nu}$.

Again, $C \leq D$ we have $c_{ij\mu} \leq d_{ij\mu}$ and $c_{ij\mu} \geq d_{ij\nu}$.

Hence, $\max\{a_{ij\mu}, c_{ij\mu}\} \leq \max\{b_{ij\mu}, d_{ij\mu}\}$ and $\min\{a_{ij\nu}, c_{ij\nu}\} \geq \min\{b_{ij\nu}, d_{ij\nu}\}$.

$$
\begin{aligned}
\text{Therefore,}\quad A+C &= \left[\left\langle \max\{a_{ij\mu}, c_{ij\mu}\}, \min\{a_{ij\nu}, c_{ij\nu}\} \right\rangle \right] \\
&\leq \left[\left\langle \max\{b_{ij\mu}, d_{ij\mu}\}, \min\{b_{ij\nu}, d_{ij\nu}\} \right\rangle \right] = B+D.
\end{aligned}
$$

Proof is similar for $A \odot C \leq B \odot D$.

3.2   Distributive lattice of GIFMs

Let A, B, $C \in G_n$ then the lattice of GIFMs $(G_n(L), \leq, +, \odot)$ is said to be distributive lattice of GIFMs if
(1) $A \odot (B+C) = (A \odot B) + (A \odot C)$ (2) $A + (B \odot C) = (A+B) \odot (A+C)$.

Theorem 4. *In a distributive lattice of GIFMs* $(G_n(L), \leq, +, \odot)$ *if A, B, $C \in$* *$G_n(L)$, $A+B = A+C$ and $A \odot B = A \odot C$, then $B = C$.*

Proof. Since, $A, B, C \in (G_n(L), \leq, +, \odot)$ we have

$$B = \left[\min\{b_{ij\mu}, \max\{a_{ij\mu}, b_{ij\mu}\}\}, \max\{b_{ij\nu}, \min\{a_{ij\nu}, b_{ij\nu}\}\} \right]$$

[By absorption property]

$$= B \odot \left[\max\{a_{ij\mu}, c_{ij\mu}\}, \min\{a_{ij\nu}, c_{ij\nu}\} \right] \text{ [Since } A+B = A+C]$$

$$= \left[\min\{b_{ij\mu}, a_{ij\mu}\}, \max\{b_{ij\nu}, a_{ij\nu}\} \right] + \left[\max\{b_{ij\mu}, c_{ij\mu}\}, \min\{b_{ij\nu}, c_{ij\nu}\} \right]$$

[By distributive law]

$$= \left[\max\{c_{ij\mu}, a_{ij\mu}\}, \min\{c_{ij\nu}, a_{ij\nu}\} \right] + \left[\min\{b_{ij\mu}, c_{ij\mu}\}, \max\{b_{ij\nu}, c_{ij\nu}\} \right]$$

[Since $B \odot A = C \odot A$]

$$= \left[\min\{c_{ij\mu}, a_{ij\mu}\}, \max\{c_{ij\nu}, a_{ij\nu}\} \right] + \left[\min\{c_{ij\mu}, b_{ij\mu}\}, \max\{c_{ij\nu}, b_{ij\nu}\} \right]$$

[By commutative law]

$$= C \odot \left[\max\{a_{ij\mu}, b_{ij\mu}\}, \min\{a_{ij\nu}, b_{ij\nu}\} \right] \text{ [By distributive law]}$$

$$= C \odot (A+C) = C \text{ [By absorption property] .}$$

Hence the theorem.

4   Generalized Intuitionistic Fuzzy Determinant (GIFD) over Distributive Lattice

The generalized intuitionistic fuzzy determinant $|A|$ of an $n \times n$ GIFM A over a distributive lattice $(G_n(L), \leq, +, \odot)$ is defined as follows:

$$det A = |A| = \sum_{\sigma \in S_n} \langle a_{1\sigma(1)\mu}, a_{1\sigma(1)\nu} \rangle \langle a_{2\sigma(2)\mu}, a_{2\sigma(2)\nu} \rangle \cdots \langle a_{n\sigma(n)\mu}, a_{n\sigma(n)\nu} \rangle$$

where S_n denotes the symmetric group of all permutations of the indices $(1, 2, \ldots, n)$.

Proposition 1. *If a GIFM B over a distributive lattice $(G_n(L), \leq, +, \odot)$, is obtained from an $n \times n$ GIFM A by multiplying the i-th row of A (i-th column) by* *$k = \langle k_1, k_2 \rangle$ such that $0 \leq k_1 + k_2 \leq 1$, then $k|A| = |B|$.*

Proof. By definition of GIFD we have

$$|B| = \sum_{\sigma \in S_n} \langle b_{1\sigma(1)\mu}, b_{1\sigma(1)v}\rangle \langle b_{2\sigma(2)\mu}, b_{2\sigma(2)v}\rangle \cdots \langle b_{n\sigma(n)\mu}, b_{n\sigma(n)v}\rangle$$

$$= \sum_{\sigma \in S_n} \langle a_{1\sigma(1)\mu}, a_{1\sigma(1)v}\rangle \langle a_{2\sigma(2)\mu}, a_{2\sigma(2)v}\rangle \cdots k\langle a_{i\sigma(i)\mu}, a_{i\sigma(i)v}\rangle \cdots \langle a_{n\sigma(n)\mu}, a_{n\sigma(n)v}\rangle$$

$$= k \sum_{\sigma \in S_n} \langle a_{1\sigma(1)\mu}, a_{1\sigma(1)v}\rangle \langle a_{2\sigma(2)\mu}, a_{2\sigma(2)v}\rangle \cdots \langle a_{i\sigma(i)\mu}, a_{i\sigma(i)v}\rangle \cdots \langle a_{n\sigma(n)\mu}, a_{n\sigma(n)v}\rangle$$

$$= k|A|.$$

Proposition 2. *Let A be an $n \times n$ GIFM over a distributive lattice $(G_n(L), \leq, +, \odot)$, if all the elements of a row (column) are $\langle 0, 1 \rangle$, then $|A| = \langle 0, 1 \rangle$.*

Proof. Since each term in $|A|$ contains a factor of each row (column) and hence contains a factor of $\langle 0, 1 \rangle$ row (column), so that each term of $|A|$ is equal to $\langle 0, 1 \rangle$ and consequently $|A| = \langle 0, 1 \rangle$.

Proposition 3. *Let A be an $n \times n$ GIFM over a distributive lattice $(G_n(L), \leq, +, \odot)$, if A is triangular, then $|A| = \prod_{i=1}^{n} \langle a_{ii\mu}, a_{iiv} \rangle$.*

Proof. Let A be a GIFM in triangular form below i.e., $\langle a_{ij\mu}, a_{ijv} \rangle = \langle 0, 1 \rangle$ for $i < j$. Now consider a term b of $|A|$

$$b = \sum_{\sigma \in S_n} \langle a_{1\sigma(1)\mu}, a_{1\sigma(1)v}\rangle \langle a_{2\sigma(2)\mu}, a_{2\sigma(2)v}\rangle \cdots \langle a_{n\sigma(n)\mu}, a_{n\sigma(n)v}\rangle.$$

Let $\sigma(1) \neq 1$, so that $1 < \sigma(1)$ and therefore $\langle a_{1\sigma(1)\mu}, a_{1\sigma(1)v} \rangle = \langle 0, 1 \rangle$ and $b = \langle 0, 1 \rangle$. This means that each term is $\langle 0, 1 \rangle$ if $\sigma(1) \neq 1$. Now let $\sigma(1) = 1$ but $\sigma(2) \neq 2$. Then $2 < \sigma(2)$ and $\langle a_{2\sigma(2)\mu}, a_{2\sigma(2)v} \rangle = \langle 0, 1 \rangle$ and $b = \langle 0, 1 \rangle$. This means that each term is $\langle 0, 1 \rangle$ if $\sigma(1) \neq 1$ and $\sigma(2) \neq 2$. In the similar manner we can prove that each term for which $\sigma(1) \neq 1$ or $\sigma(2) \neq 2 \cdots$ or $\sigma(n) \neq n$ must be $\langle 0, 1 \rangle$. Hence $|A| = \prod_{i=1}^{n} \langle a_{ii\mu}, a_{iiv} \rangle$.

4.1   Generalized intuitionistic fuzzy principal submatrix

Let $A \in (G_n(L), \leq, +, \odot)$ and $A(i_1, i_2, \cdots, j_t | j_1, j_2, \cdots, j_t)$ denote $(n-t) \times (n-t)$ submatrix obtained from A by eliminating rows $i_1, i_2, \cdots, i_t$ and columns $j_1, j_2, \cdots, j_t$ is called a principal submatrix of order $n-t$ of A.

The adjoint of an IFM over a distributive lattice is defined as below.

Definition 5. *Adjoint of an $n \times n$ GIFM A over a distributive lattice $(G_n(L), \leq, +, \odot)$, is denoted as $adjA$ and is defined by $adjA = |A_{ji}|$ where $|A_{ji}|$ is the determinant of the $(n-1) \times (n-1)$ GIFM formed by deleting row j and column i from A.*

Definition 6. *Let A be GIFM and $A \in (G_n(L), \leq, +, \odot)$, then*

$$detA = \sum_{i=1}^{n} a_{ij}A(i|j).$$

5 Generalized Intuitionistic Fuzzy Nilpotent Matrix (GIFNM) over a Distributive Lattice

If $A \in (G_n(L), \leq, +, \odot)$ and $A^m = 0$ for some $m \geq 1$, then A is called GIFNM over the distributive lattice $(G_n(L), \leq, +, \odot)$. The least positive integer m satisfying $A^m = 0$ is called the nilpotent index of A and is denoted by $h(A)$.

Definition 7. *Let A be a GIFM and $A \in G_n(L)$, then A is said to be GIFNM if and only if every principal minor of A is $\langle 0, 1 \rangle$.*

Proposition 4. *If A is an GIFNM over a distributive lattice $(G_n(L), \leq, +, \odot)$ then $detA = \langle 0, 1 \rangle$ but converse of the proposition is not true.*

Proof. Let A be an GIFNM over a distributive lattice $(G_n(L), \leq, +, \odot)$ then

$$detA = \sum_{i=1}^{n} a_{ij}A(i|j), j = 1, 2, \cdots, n.$$

Again, by the definition of the GIFNM, a GIFM A will be a nilpotent intuitionistic fuzzy matrix if and only if every principal minor of A is $\langle 0, 1 \rangle$.
Therefore $detA = \langle 0, 1 \rangle$.

Example 1. To show the converse part of the proposition, consider a GIFM over a distributive lattice $(G_n(L), \leq, +, \odot)$ as

$$A = \begin{bmatrix} \langle 0,1 \rangle & \langle 0,1 \rangle & \langle 0,1 \rangle \\ \langle 0.4,0.9 \rangle & \langle 0.5,0.8 \rangle & \langle 0.4,0.7 \rangle \\ \langle 0.5,0.6 \rangle & \langle 0.4,0.8 \rangle & \langle 0.4,0.8 \rangle \end{bmatrix}.$$

Therefore, by the Proposotion 2, we have $detA = \langle 0, 1 \rangle$ although it is not a GIFNM.

Proposition 5. *Let A be an GIFM and $A \in (G_n(L), \leq, +, \odot)$, then A is GIFNM if and only if $a_{ii}^{(k)} = \langle 0, 1 \rangle$ where $a_{ii}^{(k)}$ is the diagonal elements of A^k for all k.*

Example 2. Let us consider an GIFM over a distributive lattice $(G_n(L), \leq, +, \odot)$ as

$$A = \begin{pmatrix} \langle 0,1 \rangle & \langle 0,1 \rangle & \langle 0.4,0.7 \rangle \\ \langle 0.5,0.6 \rangle & \langle 0,1 \rangle & \langle 0.3,0.8 \rangle \\ \langle 0,1 \rangle & \langle 0,1 \rangle & \langle 0,1 \rangle \end{pmatrix}.$$

Here $a_{ii}^{(1)} = a_{ii}^{(2)} = \langle 0, 1 \rangle$ and $a_{ii}^{(3)} = \langle 0, 1 \rangle$, $i = 1, 2, 3$.

Thus GIFM A over a distributive lattice L is a GIFNM and index is $h(A) = 3$.

Proposition 6. *Let A be an GIFM over a distributive lattice L i.e., $A \in (G_n(L), \leq, +, \odot)$, then A is IFNM if and only if A is irreflexive and transitive.*

Proof. Let A be an IF irreflexive matrix i.e., $a_{ii} = \langle 0, 1 \rangle$ for all i. Since, A is IF transitive matrix we have $A^2 \leq A$ and so $A^k \leq A$ for all k. Therefore, $a_{ii}^{(k)} \leq a_{ii} = \langle 0, 1 \rangle$ for all $i, k \in N$. Therefore, A is an GIFNM.

Conversly, suppose that A is a GIFNM. If A is not a GIFNM, then $A \neq \mathbf{0}$. If $A^2 = A$ then $A = A^2 = \cdots = A^n$ and so $A^n = A \neq \mathbf{0}$. Thus a contradiction to the assumption that A is GIFNM.

Again, if $A^2 \geq A$, then $A \leq A^2 \leq \cdots \leq A^n$ and therefore $A^n \leq A \neq \mathbf{0}$, which is also a contradiction.

Hence, A must be a generalized intuitionistic fuzzy transitive matrix.

Now suppose that A is not generalized intuitionistic fuzzy irreflexive matrix. Then $a_{ii} \neq \langle 0, 1 \rangle$, for some $i \in N$. Therefore A is not a GIFNM, which is also a contradiction.

Hence A must be a generalized intuitionistic fuzzy irreflexive and transitive matrix.

Proposition 7. *Let A be a GIFM and $A \in (G_n(L), \leq, +, \odot)$, if A generalized intuitionistic fuzzy nilpotent matrix then*
(1) $A \, (adjA) = \mathbf{0}$ and $(adjA) \, A = \mathbf{0}$ (2) $(adjA)^2 = \mathbf{0}$.

Proof. Let $B = A(adjA)$. Then for any $i, j \in N$ (set of natural numbers) with $i \neq j$. We have

$$\langle b_{ij\mu}, b_{ij\nu} \rangle = \sum_{\sigma \in S_n} \langle a_{1\sigma(1)\mu}, a_{1\sigma(1)\mu} \rangle \cdots \langle a_{i\sigma(i)\mu}, a_{i\sigma(i)\nu} \rangle \cdots \langle a_{n\sigma(n)}, a_{i\sigma(i)\nu} \rangle.$$

Let $\sigma \in S_n$ be an arbitrary,
Case-(i): $\sigma^l(i) \neq j$ for $l \geq 1$.
Then there exists d such that $1 \leq d \leq n$, $\sigma^l(i) = i$ and $i, \sigma^l(i) \cdots \sigma^{d-1}(i)$ are mutually different and belong to N. Then
$$\langle a_{1\sigma(1)\mu}, a_{1\sigma(1)\nu} \rangle \cdots \langle a_{i\sigma(i)\mu}, a_{i\sigma(i)\nu} \rangle \cdots \langle a_{n\sigma(n)\mu}, a_{n\sigma(n)\nu} \rangle$$
$$\leq \langle a_{i\sigma(i)\mu}, a_{i\sigma(i)\nu} \rangle \cdots \langle a_{\sigma(i)\sigma^2(i)\mu}, a_{\sigma(i)\sigma^2(i)\nu} \rangle \cdots \langle a_{\sigma^{d-1}(i)i\mu}, a_{\sigma^{d-1}(i)i\nu} \rangle$$
$$\leq (A^d)_{ii} \text{ [since for an IFNM } a_{ii}^k = \langle 0, 1 \rangle \text{ for all } i, k \in N \text{ where } A^k = \left[a_{ii}^k \right]$$
$$= \left[\langle 0, 1 \rangle \right].$$
Case-(ii): There exists l such that $\sigma^l(i) = j$.
Then there exists d such that $1 \leq d \leq n$, $\sigma^l(i) = j$ and $i, \sigma^l(j) \cdots \sigma^{d-1}(j)$ are mutually different and belong to N. Then
$$\langle a_{1\sigma(1)\mu}, a_{1\sigma(1)\nu} \rangle \cdots \langle a_{i\sigma(i)\mu}, a_{i\sigma(i)\nu} \rangle \cdots \langle a_{i\sigma(j)\mu}, a_{i\sigma(j)\nu} \rangle \cdots \langle a_{n\sigma(n)\mu}, a_{n\sigma(n)\nu} \rangle$$
$$\leq \langle a_{i\sigma(j)\mu}, a_{i\sigma(j)\nu} \rangle \cdots \langle a_{\sigma(j)\sigma^2(j)\mu}, a_{\sigma(j)\sigma^2(j)\nu} \rangle \cdots \langle a_{\sigma^{d-1}(j)i\mu}, a_{\sigma^{d-1}(j)i\nu} \rangle$$

$\leq (A^d)_{ii}$ [since for an IFNM $a_{ii}^k = \langle 0,1 \rangle$ for all $i,k \in N$ where $A^k = \left[a_{ii}^k \right]$
$= \left[\langle 0,1 \rangle \right]$.

Therefore, for any $i, j \in N$ with $i \neq j$, we have

$$\langle b_{ij\mu}, b_{ijv} \rangle = \sum_{\sigma \in s_n} \langle a_{1\sigma(1)\mu}, a_{1\sigma(1)\mu} \rangle \cdots \langle a_{i\sigma(i)\mu}, a_{i\sigma(i)v} \rangle \cdots \langle a_{n\sigma(n)}, a_{i\sigma(i)v} \rangle = \langle 0,1 \rangle.$$

If $i = j$, it is clear that $b_{ii} = \langle 0,1 \rangle$. Thus $B = A(adjA) = \mathbf{0}$.

Similarly, it can be prove that $(adjA)A = \mathbf{0}$.

Proposition 8. *Let A be a GIFNM over a distributive lattice L i.e., $A \in F_n(L)$, then $h(A) = 3$ if and only if $AA^T = \mathbf{0}$ and for some $i, j \in N$, $R_i \wedge R_j^T \neq \langle 0,1 \rangle$, where R_i and R_j are the i-th and j-th rows of the IFNM A.*

Proof. Let the index of a GIFM A be 3 i.e., $h(A) = 3$. Since $A^2 \neq \mathbf{0}$ then there must exists some rows say s-th and t-th such that $R_i \wedge R_j^T \neq \langle 0,1 \rangle$. Now we show that $AA^T = \mathbf{0}$. Suppose $AA^T \neq \mathbf{0}$, then we can find p, q such that $\langle a_{pq\mu}, a_{pqv} \rangle \langle a_{qp\mu}, a_{qpv} \rangle > \langle 0,1 \rangle$. Therefore $\langle a_{pq\mu}, a_{pqv} \rangle \langle a_{qp\mu}, a_{qpv} \rangle \langle a_{pq\mu}, a_{pqv} \rangle > \langle 0,1 \rangle$, which is a term of (p,q)-th entry of the matrix A^3. So $\sum_{1 \leq i_1, i_2 \leq n} \langle a_{qi_1\mu}, a_{qi_1v} \rangle$

$\langle a_{i_1i_2\mu}, a_{i_1i_2v} \rangle \langle a_{i_2p\mu}, a_{i_2pv} \rangle > \langle a_{qp\mu}, a_{qpv} \rangle \langle a_{pq\mu}, a_{pqv} \rangle \langle a_{qp\mu}, a_{qpv} \rangle > \langle 0,1 \rangle$, which leads to a contradiction, since $A^3 = \mathbf{0}$.

Conversely, $R_i \wedge R_j^T \neq \langle 0,1 \rangle$ for some rows say i, j, then $A^2 \neq \mathbf{0}$. Therefore, from $AA^T = \mathbf{0}$ we get $A^3 = \mathbf{0}$. Hence A is an IFNM of index 3.

Note 1. Let A be an GIFNM over a distributive lattice L i.e., $A \in G_n(L)$, and $AA^T = \mathbf{0}$ and for all $i, j \in N$, $R_i \wedge R_j^T = \langle 0,1 \rangle$ then A is neither GIFNM and nor converge to a GIFM.

References

1. K. Atanassov, Intuitionistic fuzzy sets, Fuzzy Sets and Systems, 20 (1986), 87–96.
2. M.Z. Ragab and E.G. Emam, The determinant and adjoint of a square fuzzy matrix, Fuzzy Sets and Systems, 61 (1994), 297–307.
3. M. Pal, Intuitionistic fuzzy determinant, V.U.J.Physical Sciences, 7 (2001), 87–93.
4. M. Pal, S.K. Khan and A.K. Shyamal, Intuitionistic fuzzy matrices, Notes on Intuitionistic Fuzzy Sets, 8(2) (2002), 51–62.
5. T.K. Mondal and S.K. Samanta, Generalized intuitionistic fuzzy sets, The Journal of Fuzzy Mathematics, 10(4) (2002,) 839–862.
6. M. Bhowmik and M. Pal, Generalized intuitionistic fuzzy matrices, Far-East Journal of Mathematical Sciences, 29(3) (2008), 533–554.
7. Y. Give'on, Lattice matrices, Information Control, 7(3) (1964), 477–484.

On Quartiles of Generalised Fuzzy Number with Different Left Height and Right Height

Rituparna Chutia[*,†], Supahi Mahanta[‡], D. Datta[§]

†Department of Mathematics, Gauhati University, Assam – 781 014, India.
‡Department of Statistics, Gauhati University, Assam – 781 014, India.
§Health Physics Division, Bhabha Atomic Research Centre, Mumbai – 400 085, India.

Abstract. Quartiles of generalised fuzzy numbers with different left height and right height are investigated. These quartiles of the generalised fuzzy number are evaluated on the basis of definitions put forwarded to define distribution of cardinality of those generalised fuzzy number. Locations of these quartiles are investigated on the basis of the distribution of cardinality of the generalised fuzzy number.

Key words: Quartiles, generalised fuzzy number, cardinality.

1 Introduction

Engineering and management models data are often encountered by imprecision, information lacking and inaccurate expert judgement. In such a situation fuzzy numbers are best tools to handle such type of problems. With the advances in fuzzy mathematics different types of fuzzy numbers are proposed in different times. Recently, a generalised fuzzy number (GFN) with different left height and right height and their arithmetic are proposed in [1]. And it very essential to study the quartiles of GFN for an application viewpoint, so that these scalar approximation can be used in statistical analysis.

Some of developments in theory and applications of central tendency of fuzzy numbers are discussed in this paragraph. Dubois and Prade [2] introduced the mean value of a fuzzy number as a closed interval bounded by the expectations calculated from its upper and lower distribution functions. Interval-valued possibilistic mean of fuzzy numbers of fuzzy number is introduced by Carlsson and Fuller [3] and investigated its relationship to interval-valued probabilistic mean. Further, weighted possibilistic mean, variance and covariance of fuzzy numbers are introduced by Fuller and Majlender [4]. Bodjanova [5] also introduced the median value and median interval of fuzzy numbers. In his work, median value and median interval are evaluated with special attention towards trapezoidal fuzzy numbers and their modifications by concentration and dilation. Chen and Tan [6] introduced the possibilistic mean, variance and covariance of the multiplication of fuzzy numbers and some related properties are

---

* Email: Rituparnachutia7@rediffmail.com

explored. More recent work on central tendency of fuzzy number is found in [7] where the notion of median of an random fuzzy number on the basis of an L^1-type metric between fuzzy numbers are explored.

However, few work are available on evaluation of quartiles of GFN with different left height and right height is one of the motivation to this present work. Also, the evaluations of quartiles and their locations of the GFN with different left height and right height is being motivated by the work of [5] where mean value and the mean interval of a fuzzy number are evaluated. It was clearly mentioned that these work can further be extended to any q^{th} percentile, $q \in (0, 100]$.

2   Definitions and Notations

In this section, brief review of some concepts of generalised fuzzy number with different left height and right height are put forwarded.

2.1   Generalised fuzzy number

Let $\tilde{A}$ is represented by $\tilde{A} = (a, b, c, d; h_L, h_R)$ on the real line IR is called a GFN with different left height and right height, where a, b, c and d real values, h_L is called the left height of the GFN $\tilde{A}$, h_R is called the right height of the GFN, $h_L \in [0, 1]$ and $h_R \in [0, 1]$ [1]. For now, let $F(IR)$ be the set of all GFN with different left height and right height. If $h_L = h_R = 1$ then the GFN reduces to a standard trapezoidal fuzzy number.

The membership function of GFN $\tilde{A}$ with different left height and right height is as given below

$$\mu_{\tilde{A}}(x) = \begin{cases} \frac{h_L(x-a)}{b-a}, & \text{if } a \leq x \leq b; \\ \frac{h_L(c-b)+(h_R-h_L)(x-b)}{c-b}, & \text{if } b \leq x \leq c; \\ \frac{h_R(x-d)}{c-d}, & \text{if } c \leq x \leq d; \\ 0, & \text{otherwise} \end{cases} \tag{1}$$

For the GFN $\tilde{A}$ described by the membership function (1) has the following α-cuts:

$$\tilde{A}_\alpha = \begin{cases} \left[\frac{ah_L+\alpha(b-a)}{h_L}, \frac{dh_L+\alpha(c-d)}{h_H} \right], & \text{if } 0 \leq \alpha \leq h_L; \\ \left[\frac{b(h_H-h_L)+(\alpha-h_L)(c-b)}{h_H-h_L}, \frac{dh_L+\alpha(c-d)}{h_H} \right], & \text{if } h_L \leq \alpha \leq h_H; \end{cases} \tag{2}$$

if $(h_L < h_H)$, and

$$\tilde{A}_\alpha = \begin{cases} \left[\frac{ah_L+\alpha(b-a)}{h_L}, \frac{dh_L+\alpha(c-d)}{h_H} \right], & \text{if } 0 \leq \alpha \leq h_H; \\ \left[\frac{ah_L+\alpha(b-a)}{h_L}, \frac{b(h_H-h_L)+(\alpha-h_L)(c-b)}{h_H-h_L} \right], & \text{if } h_H \leq \alpha \leq h_L; \end{cases} \tag{3}$$

if $(h_L > h_H)$.

Cardinality of a GFN $\tilde{A}$ described by (1) is the value of the area under the membership function. It is given by the integral

$$
\begin{aligned}
Card\tilde{A} &= \int_c^d \mu_{\tilde{A}}(x)dx \\
&= \int_0^{h_L} \left(\frac{dh_L + \alpha(c-d)}{h_H} - \frac{ah_L + \alpha(b-a)}{h_L} \right) d\alpha \\
&\quad + \int_{h_L}^{h_H} \left(\frac{dh_L + \alpha(c-d)}{h_H} - \frac{b(h_H - h_L) + (\alpha - h_L)(c-b)}{h_H - h_L} \right) d\alpha \\
&= \frac{h_H(d-b) + h_L(c-a)}{2}
\end{aligned}
\tag{4}
$$

Cardinality summarised in (4); is obtained on consideration of α-cuts (2). However, (4) can be obtained on consideration of α-cuts (3) also.

3   Quartiles of GFN with Different Left Height and Right Height

The quartiles of a fuzzy number and different properties of them are discussed and put forwarded in the form of proposition. The definition of 2nd quartiles or the median value of a fuzzy number $\tilde{A}$ is a possible scalar representative value of $\tilde{A}$. Here some of the descriptions of the 2nd quartile are adopted from [5]. Definitions and some descriptions related to 1st and 3rd quartile are also put forwarded. 1st, 2nd and 3rd quartiles are real values and are represented as m_{25}, m_{50} and m_{75} respectively, from now.

Definition 1 *([5]). The median value or 2nd quartile of a GFN $\tilde{A}$ with different left and right height given by (1) is a real number m_{50} such that*

$$
\int_a^{m_{50}} \mu_{\tilde{A}}(x)dx = \int_{m_{50}}^d \mu_{\tilde{A}}(x)dx
\tag{5}
$$

For practical purpose the above expression (5) can be rewritten as

$$
\int_a^{m_{50}} \mu_{\tilde{A}}(x)dx = \frac{1}{2} Card\tilde{A}
\tag{6}
$$

Definition 2 *The 1st quartile of a GFN $\tilde{A}$ with different left and right height given by (1) is a real number m_{25} such that*

$$
\int_a^{m_{25}} \mu_{\tilde{A}}(x)dx + \int_{m_{25}}^{m_{50}} \mu_{\tilde{A}}(x)dx = \int_{m_{50}}^{m_{75}} \mu_{\tilde{A}}(x)dx + \int_{m_{75}}^d \mu_{\tilde{A}}(x)dx
\tag{7}
$$

For practical purpose the above expression (7) can be rewritten as

$$
\int_a^{m_{25}} \mu_{\tilde{A}}(x)dx = \frac{1}{4} Card\tilde{A}
\tag{8}
$$

Definition 3 *The 3rd quartile of a GFN $\tilde{A}$ with different left and right height given by (1) is a real number m_{75} such that*

$$\int_a^{m_{75}} \mu_{\tilde{A}}(x)dx = \frac{3}{4}Card\tilde{A} \tag{9}$$

Depending upon the distribution of the cardinality of the GFN $\tilde{A}$ with different left height and right height, following definitions are put forwarded:

1. A GFN is a GFN with light tails if $max\{\int_a^b \mu_{\tilde{A}}(x)dx, \int_c^d \mu_{\tilde{A}}(x)dx\} \leq \frac{1}{2}\int_a^d \mu_{\tilde{A}}(x)dx$.
2. A GFN is a GFN with very light tails if $max\{\int_a^b \mu_{\tilde{A}}(x)dx, \int_c^d \mu_{\tilde{A}}(x)dx\} \leq \frac{1}{4}\int_a^d \mu_{\tilde{A}}(x)dx$.
3. A GFN is a GFN with heavy left tail if $\int_a^b \mu_{\tilde{A}}(x)dx > \frac{1}{2}\int_a^d \mu_{\tilde{A}}(x)dx$.
4. A GFN is a GFN with very heavy left tail if $\int_a^b \mu_{\tilde{A}}(x)dx > \frac{3}{4}\int_a^d \mu_{\tilde{A}}(x)dx$.
5. A GFN is a GFN with heavy right tail if $\int_c^d \mu_{\tilde{A}}(x)dx > \frac{1}{2}\int_a^d \mu_{\tilde{A}}(x)dx$.
6. A GFN is a GFN with very heavy right tail if $\int_c^d \mu_{\tilde{A}}(x)dx > \frac{3}{4}\int_a^d \mu_{\tilde{A}}(x)dx$.

Proposition 1 *Let $\tilde{A} = (a,b,c,d;h_L,h_H)$ be a GFN with a light tails then*

$$m_{50} = \frac{b(h_L+h_H) - 2h_L c + \sqrt{2(c-b)\{h_L^2(c-a) + h_H^2(d-b) + h_L h_H(c+a-b-d)\}}}{h_H - h_L} \tag{10}$$

Proof. Since $max\{\int_a^b \mu_{\tilde{A}}(x)dx, \int_c^d \mu_{\tilde{A}}(x)dx\} \leq \frac{1}{2}\int_a^d \mu_{\tilde{A}}(x)dx$, the median value or 2nd quartile m_{50} must be in the interval $[b,c]$. Then,

$$\frac{1}{2}Card\tilde{A} = \int_a^b \mu_{\tilde{A}}(x)dx + \int_b^{m_{50}} \mu_{\tilde{A}}(x)dx$$

Hence the proof follows immediately.

Proposition 2 *Let $\tilde{A} = (a,b,c,d;h_L,h_H)$ be a GFN. Then*

$$m_{50} = a + \sqrt{\frac{(b-a)Card\tilde{A}}{h_L}} \tag{11}$$

if $\tilde{A}$ has a heavy or very heavy left tail, and

$$m_{50} = d - \sqrt{\frac{(d-c)Card\tilde{A}}{h_H}} \tag{12}$$

if $\tilde{A}$ has a heavy or very heavy right tail.

Proof. The proof follows immediately from the above definitions.

Proposition 3 *Let $\tilde{A} = (a,b,c,d;h_L,h_H)$ be a GFN. Then*

$$m_{25} = a + \sqrt{\frac{(b-a)Card\tilde{A}}{2h_L}} \tag{13}$$

if $\tilde{A}$ has a heavy or very heavy left tail, and

$$m_{25} = d - \sqrt{\frac{(d-c)Card\tilde{A}}{2h_H}} \tag{14}$$

if $\tilde{A}$ has a heavy or very heavy right tail.

Proof. The proof follows immediately from the above definitions.

Proposition 4 *Let $\tilde{A} = (a,b,c,d;h_L,h_H)$ be a GFN with a very heavy tails. Then*

$$m_{75} = a + \sqrt{\frac{3(b-a)Card\tilde{A}}{2h_L}} \tag{15}$$

if $\tilde{A}$ has a very heavy left tail, and

$$m_{75} = d - \sqrt{\frac{3(d-c)Card\tilde{A}}{2h_H}} \tag{16}$$

if $\tilde{A}$ has a very heavy right tail.

Proof. The proof follows immediately from the above definitions.

Proposition 5 *Let $\tilde{A} = (a,b,c,d;h_L,h_H)$ be a GFN with a heavy left tail. Then*

$$m_{75} = b \tag{17}$$

if $2h_L(b-a) = 3Card\tilde{A}$, and

$$m_{75} = \frac{2(bh_H - ch_L) + \sqrt{(c-b)\left[4h_L^2(c-b) - (h_H - h_L)\{h_L(7b - 4a - 3d) - 3h_H(c-a)\}\right]}}{2(h_H - h_L)} \tag{18}$$

if $2h_L(b-a) < 3Card\tilde{A}$.

Proof. It is assumed that $\tilde{A}$ has a heavy left tail such that $\int_a^b \mu_{\tilde{A}}(x)dx > \frac{1}{2}Card\tilde{A}$. Therefore $\frac{1}{2}Card\tilde{A} < \int_a^b \mu_{\tilde{A}}(x)dx \leq \frac{3}{4}Card\tilde{A}$ which implies $2h_L(b-a) \geq 3Card\tilde{A}$.

Suppose that $\int_a^b \mu_{\tilde{A}}(x)dx < \frac{3}{4}Card\tilde{A}$ this implies $2h_L(b-a) < 3Card\tilde{A}$. Then $b < m_{75} < c$ and

$$\int_a^b \mu_{\tilde{A}}(x)dx + \int_b^{m_{75}} \mu_{\tilde{A}}(x)dx = \frac{3Card\tilde{A}}{4}$$

$$m_{75} = \frac{2(bh_H - ch_L) + \sqrt{(c-b)\left[4h_L^2(c-b) - (h_H - h_L)\{h_L(7b - 4a - 3d) - 3h_H(c-a)\}\right]}}{2(h_H - h_L)}$$

which yields (18). Eq. (17) can be proved easily from Eq. (15).

Proposition 6 *Let $\tilde{A} = (a,b,c,d;h_L,h_H)$ be a GFN with a heavy right tail. Then*

$$m_{75} = c \tag{19}$$

if $2h_H(d-c) = 3Card\tilde{A}$, and

$$m_{75} = \frac{2h_L(b-2c) + 2bh_H + \sqrt{2(c-b)\left[8h_L^2(c-b) - (h_H - h_L)\{h_L(4b - c - 3a) + h_H(b-d)\}\right]}}{2(h_H - h_L)} \tag{20}$$

if $2h_H(d-c) < 3Card\tilde{A}$.

Proof. It is assumed that $\tilde{A}$ has a heavy right tail such that $\int_c^d \mu_{\tilde{A}}(x)dx > \frac{1}{2}Card\tilde{A}$. Therefore, $\frac{1}{2}Card\tilde{A} < \int_c^d \mu_{\tilde{A}}(x)dx \leq \frac{3}{4}Card\tilde{A}$ which implies $2h_H(d-c) \geq 3Card\tilde{A}$. Suppose that $\int_c^d \mu_{\tilde{A}}(x)dx < \frac{3}{4}Card\tilde{A}$ this implies $2h_H(d-c) < 3Card\tilde{A}$. Then $b < m_{75} < c$ and

$$\int_{m_{75}}^c \mu_{\tilde{A}}(x)dx + \int_c^d \mu_{\tilde{A}}(x)dx = \frac{3Card\tilde{A}}{4}$$

$$\int_a^b \mu_{\tilde{A}}(x)dx + \int_b^{m_{75}} \mu_{\tilde{A}}(x)dx = \frac{Card\tilde{A}}{4}$$

$$m_{75} = \frac{2h_L(b-2c) + 2bh_H + \sqrt{2(c-b)[8h_L^2(c-b) - (h_H - h_L)\{h_L(4b - c - 3a) + h_H(b-d)\}]}}{2(h_H - h_L)}$$

which proves (20). Eq. (19) follows immediately from Eq. (16).

Proposition 7 *Let $\tilde{A} = (a,b,c,d;h_L,h_H)$ be a GFN with a heavy or very heavy tails. Then*

$$\sqrt{\frac{h_L}{2}} < \mu_{\tilde{A}}(m_{50}) = \sqrt{\frac{h_L Card\tilde{A}}{b-a}} < h_L \tag{21}$$

if $\tilde{A}$ has a heavy or very heavy left tail, and

$$\sqrt{\frac{h_H}{2}} < \mu_{\tilde{A}}(m_{50}) = \sqrt{\frac{h_H Card\tilde{A}}{d-c}} < h_H \tag{22}$$

if $\tilde{A}$ has a heavy or very heavy right tail.

Proof. This prove can be done easily.

Proposition 8 *Let $\tilde{A} = (a,b,c,d;h_L,h_H)$ be a GFN with a heavy or very heavy tails. Then*

$$\sqrt{\frac{h_L}{4}} < \mu_{\tilde{A}}(m_{25}) = \sqrt{\frac{h_L Card\tilde{A}}{2(b-a)}} < h_L \tag{23}$$

if $\tilde{A}$ has a heavy or very heavy left tail, and

$$\sqrt{\frac{h_H}{4}} < \mu_{\tilde{A}}(m_{25}) = \sqrt{\frac{h_H Card\tilde{A}}{2(d-c)}} < h_H \tag{24}$$

if $\tilde{A}$ has a heavy or very heavy right tail.

Proof. These can be proved easily as done for (21) and (22).

Proposition 9 *Let $\tilde{A} = (a,b,c,d;h_L,h_H)$ be a GFN with a very heavy tails. Then*

$$\sqrt{\frac{3h_L}{2}} < \mu_{\tilde{A}}(m_{75}) = \sqrt{\frac{3h_L Card\tilde{A}}{2(b-a)}} < h_L \tag{25}$$

if $\tilde{A}$ has a very heavy left tail, and

$$\sqrt{\frac{3h_H}{2}} < \mu_{\tilde{A}}(m_{75}) = \sqrt{\frac{3h_H Card\tilde{A}}{2(d-c)}} < h_H \tag{26}$$

if $\tilde{A}$ has a very heavy right tail.

Proof. These can be proved easily as done for (21) and (22).

4   Conclusions

Quartiles of GFN with different left height and right height are are formulated. Also the locations of these 1st, 2nd and 3rd quartiles and its membership grade could be found from the formulation. Statistical properties such as quartile coefficient of skewness and quartile coefficient of dispersion are also could be evaluated from the derived quartiles. These information would be helpful in determining the nature of the tails of the GFN with different left height and right height. Also other quartile properties could be also evaluated from those derived quartiles.

References

1. S. M. Chen, A. Munif, G. S. Chen, H. C. Liu, B. C. Kuo, Fuzzy risk analysis based on ranking generalised fuzzy numbers with different left heights and right heights, Expert Systems with Applications, 39 (2012), 6320–6334.
2. D. Dubois, H. Parde, The mean value of a fuzzy number, Fuzzy Sets and Systems, 24 (1987), 179–300.
3. C. Carlsson C, R. Fuller On possibilistic mean value and variance of fuzzy numbers, Fuzzy Sets and Systems, 122 (2001), 315–326.
4. R. Fuller, P. Majlender, On weighted possibilistic mean and variance of fuzzy numbers, Fuzzy Sets and Systems, 136 (2003), 203–215.
5. S. Bodjanova, Median value and median interval of a fuzzy number, Information Sciences, 172 (2005), 73–89.
6. W. Chen, S. Tan, On the possibilistic mean value and variance of multiplication of fuzzy numbers, Journal of Computational and Applied Mathematics, 232 (2009), 327–334.
7. B. Sinova, M. A. Gil, A. Colubi, S. V. Aelst, The median of a random fuzzy number. The 1-norm distance approach, Fuzzy Sets and Systems, 200 (2012), 99–115.

Fuzzification of Pollutant Concentration under the Gaussian Plume Model

Supahi Mahanta[*,†], Rituparna Chutia[**,‡] and Hemanta K. Baruah[***,†]

†Department of Statistics, Gauhati University, Guwahati – 14, Assam, India.
‡Department of Mathematics, Gauhati University, Guwahati – 14, Assam, India.

Abstract. This article describes the generalized form of the membership functions of pollutant concentration of atmospheric dispersion defined by the Gaussian Plume model. The membership functions of concentration for different stability categories of atmosphere have been derived using Lagrange's method of interpolation. Assuming that the parameters concerned are triangular, and based on the Randomness Fuzziness Consistency Principle the probability density functions associated with these membership functions of pollutant concentration have also been obtained.

Key words: Gaussian plume model, Lagrange's method of interpolation, randomness-fuzziness consistency principle.

1 Introduction

The term "Atmospheric Dispersion" means spread to the environment through the atmosphere. It consists of two components, viz, the adjective transport by wind in the downwind direction and the spread due to turbulent diffusion in the vertical and in the horizontal directions. Atmospheric dispersion models are mathematical expressions or algorithms relating the quantity of a pollutant released to the atmosphere to its concentration at a given location. The concentration of an air pollutant at a given place is a function of a number of variables, including the emission rate, the distance of the receptor from the source and the atmospheric conditions. The most important atmospheric conditions are wind speed, wind direction, and the vertical temperature structure of the local atmosphere [1]. Atmospheric dispersion is a topic in meteorology especially in branches commonly known as micro-meteorology and air pollution meteorology, because of its extensive applications in air pollution evaluation and control. The distribution of the concentration along the crosswind direction was found to fit closely a Gaussian distribution. The plume-spread parameter could then be expressed in terms of the standard deviation of the Gaussian distribution [2].

* Corresponding author. Email: supahi_mahanta@rediffmail.com
** Email: Rituparnachutia7@rediffmail.com
*** Email: hemanta_bh@yahoo.com

2 Gaussian Plume Model

The air quality dispersion model considered here is the classical Gaussian plume model. This model assumes a Gaussian distribution of concentration along the co-ordinate directions. The Gaussian plume model [2] is the most widely used method of estimating downwind concentration of airborne material released to the atmosphere. Sutton [3, 4] derived an air pollutant plume dispersion equation which included the assumption of Gaussian distribution for the vertical and the crosswind dispersion of the plume, and also included the effect of ground reflection of the plume. Input parameters of the standard Gaussian Plume model considered are: wind speed, source height, horizontal and vertical standard deviations, quantity of the contaminant and stability category of weather.

The Gaussian plume model that provides the time integrated air concentration at any downwind distance is given by

$$C(x,y,z) = \frac{Q}{2\pi\sigma_y\sigma_z}\exp(-\frac{y^2}{2\sigma_y^2})[\exp(-\frac{(z-H)^2}{2\sigma_z^2}) + \exp(-\frac{(z+H)^2}{2\sigma_z^2})]$$

where $C(x,y,z)$ is the concentration of the pollutant (in micrograms per cubic metre) at any point x metres downwind of the source, y metres crosswind from the emission plume centreline and z metres above ground level, Q is the quantity or mass of the pollutant (in grams per seconds), u is the average wind speed (in metres per second), σ_z is standard deviation in the vertical direction (in metre), σ_y is the standard deviation in the crosswind direction (in metre), H is the effective stack height of the source above ground level (in metres).

The values of horizontal and vertical dispersion co-efficients (σ_y and σ_z) here can be seen to be

$$\sigma_y = a_y x^{0.9071}$$
$$\sigma_z = a_z x^{b_z} + c_z$$

where the co-efficient a_y, a_z, b_z and c_z can be obtained from the table of parameters for Pasquill-Gifford σ_y and σ_z [5]. Based on the temperature gradient, atmospheric conditions are categorized into six classes, so called the Pasquill stability classes [6], labelled A through F. Classes A through C are unstable conditions, class D is neutral, and classes E and F are stable.

The purpose of this work is to evaluate the ground level concentrations throughout the model domain. The ground level concentrations directly downwind are of interest, since pollution concentration will be the highest along that axis. The ground level concentrations along the centreline of the plume are obtained by setting $y = 0$ in the model. The expression for the centreline ground level concentration becomes

$$C(x,0,0) = \frac{Q}{\pi\sigma_y\sigma_z}\exp(-\frac{H^2}{2\sigma_z^2})$$

Gaussian plume models are applicable for downwind distance, $x > 100$ m, because near the source concentration approaches infinity [7]. Plume rise $\triangle h$ (in meter) plays an important role in determining the ground level concentrations. The plume rise is added to the height of the plume's source point to obtain the effective stack height H. The plume rise equation due to Moses and Carson [8] is as follows:

$$\triangle h = A\frac{w_0 d}{u} + B\frac{\sqrt{Q_H}}{u}$$

where $Q_H = r\pi w_0 C_p(T_s - T_a)$ is the heat release rate (Cal/sec) and d is the stack diameter (m), r is the stack radius (m), w_0 is the discharge velocity (m/sec), C_p is the specific heat at constant pressure (Cal/kg-K), T_a is the ambient temperature (K) and T_s is the gas exit temperature (K). The parameters A and B depends upon the stability classes and are given by $A = 3.47, B = 0.333$ for unstable, $A = 0.35, B = 0.17$ for neutral and $A = -1.04, B = 0.17$ for stable stability classes [2].

3 Methodology

This article demonstrates the general form of membership function of atmospheric dispersion defined by the Gaussian Plume Model, when the input parameters of the model are triangular fuzzy in nature. The essence of this work is based on a result linking fuzziness with randomness defined by Baruah [9–12] based on superimposition of sets [13]. The methodology used for evaluating the membership functions of atmospheric dispersion defined by the Gaussian Plume Model is based on Lagrange's method of interpolation on discretized values of the α cuts of concentration. These membership functions give two probability laws based on the *randomness-fuzziness consistency principle* [9, 14, 15]. According to this principle the Dubois-Prade left reference function of a membership function is a distribution function in the measure theoretic sense, and similarly the Dubois-Prade right reference function is a complementary distribution function.

4 Fuzzified Pollutant Concentration of the Model

The membership functions and membership curves of the ground level concentration along the plume centreline for different stability categories are evaluated. The input parameters wind speed, discharge velocity, ambient temperature and gas exit temperature are considered as triangular fuzzy variables. The evaluation of the model output along the downwind distances is approximated by the Lagrange's polynomial of degree four on discretized values of the α-cuts of $C(x,0,0)$.

Let the triangular fuzzy numbers for the uncertain parameters are
$u = [a_1, a_2, a_3]$, $T_s = [b_1, b_2, b_3]$, $T_a = [c_1, c_2, c_3]$ and $w_0 = [d_1, d_2, d_3]$.
The generalized form of α - cuts for the concentration $C(x, 0, 0)$ is given as

$$
\alpha = \begin{cases}
\dfrac{Qe^{-\frac{1}{2}\left[\frac{h[a_1+(a_2-a_1)\alpha]+Ad[d_3-(d_3-d_2)\alpha]+Br\sqrt{\pi C_p[d_3-(d_3-d_2)\alpha][(b_3-c_1)-(b_3-b_2+c_2-c_1)\alpha]}}{[a_1+(a_2-a_1)\alpha](a_zx^{b_z}+c_z)}\right]^2}}{\pi a_y x^{0.9071}(a_zx^{b_z}+c_z)[a_3-(a_3-a_2)\alpha]}, \\[2em]
\dfrac{Qe^{-\frac{1}{2}\left[\frac{h[a_3-(a_3-a_2)\alpha]+Ad[d_1+(d_2-d_1)\alpha]+Br\sqrt{\pi C_p[d_1+(d_2-d_1)\alpha][(b_1-c_3)+(b_2-b_1+c_3-c_2)\alpha]}}{[a_3-(a_3-a_2)\alpha](a_zx^{b_z}+c_z)}\right]^2}}{\pi a_y x^{0.9071}(a_zx^{b_z}+c_z)[a_1+(a_2-a_1)\alpha]}
\end{cases}
$$

where the co-efficient a_y, a_z, b_z and c_z can be obtained from the table of parameters for Pasquill-Gifford σ_y and σ_z dependent on downwind distances x and different stability categories [5].

Say, the lower α-cut of $C(x, 0, 0)$ is

$$
X = \frac{Qe^{-\frac{1}{2}\left[\frac{h[a_1+(a_2-a_1)\alpha]+Ad[d_3-(d_3-d_2)\alpha]+Br\sqrt{\pi C_p[d_3-(d_3-d_2)\alpha][(b_3-c_1)-(b_3-b_2+c_2-c_1)\alpha]}}{[a_1+(a_2-a_1)\alpha](a_zx^{b_z}+c_z)}\right]^2}}{\pi a_y x^{0.9071}\left(a_zx^{b_z}+c_z\right)[a_3-(a_3-a_2)\alpha]}
$$

Then for five different values of α the values of lower α-cuts are,

α	Values of X
0	$X\vert_{\alpha=0} = (M_1, \text{say})$
0.25	$X\vert_{\alpha=0.25} = (M_2, \text{say})$
0.50	$X\vert_{\alpha=0.50} = (M_3, \text{say})$
0.75	$X\vert_{\alpha=0.75} = (M_4, \text{say})$
1	$X\vert_{\alpha=1} = (M_5, \text{say})$

Thus the left reference function obtained using Lagrangian polynomial approximation as,

$$
\begin{aligned}
F(X) = &\frac{(X-M_1)(X-M_3)(X-M_4)(X-M_5)0.25}{(M_2-M_1)(M_2-M_3)(M_2-M_4)(M_2-M_5)} + \frac{(X-M_1)(X-M_2)(X-M_4)(X-M_5)0.50}{(M_3-M_1)(M_3-M_2)(M_3-M_4)(M_3-M_5)} \\
&+ \frac{(X-M_1)(X-M_2)(X-M_3)(X-M_5)0.75}{(M_4-M_1)(M_4-M_2)(M_4-M_3)(M_4-M_5)} + \frac{(X-M_1)(X-M_2)(X-M_3)(X-M_4)}{(M_5-M_1)(M_5-M_2)(M_5-M_3)(M_5-M_4)},
\end{aligned}
$$

$$
\frac{Qe^{-\frac{1}{2}\left[\frac{(ha_1+Add_3)+Br\sqrt{\pi C_pd_3(b_3-c_1)}}{a_1(a_zx^{b_z}+c_z)}\right]^2}}{\pi a_3 a_y x^{0.9071}(a_zx^{b_z}+c_z)} \leq X \leq \frac{Qe^{-\frac{1}{2}\left[\frac{(ha_2+Add_2)+Br\sqrt{\pi C_pd_2(b_2-c_2)}}{a_2(a_zx^{b_z}+c_z)}\right]^2}}{\pi a_2 a_y x^{0.9071}(a_zx^{b_z}+c_z)}
$$

Again, let the upper α-cut of $C(x, 0, 0)$ be

$$
X = \frac{Qe^{-\frac{1}{2}\left[\frac{h[a_3-(a_3-a_2)\alpha]+Ad[d_1+(d_2-d_1)\alpha]+Br\sqrt{\pi C_p[d_1+(d_2-d_1)\alpha][(b_1-c_3)+(b_2-b_1+c_3-c_2)\alpha]}}{[a_3-(a_3-a_2)\alpha](a_zx^{b_z}+c_z)}\right]^2}}{\pi a_y x^{0.9071}\left(a_zx^{b_z}+c_z\right)[a_1+(a_2-a_1)\alpha]}
$$

Then for five different values of α the values of upper α-cuts are,

α	Values of X
0	$X\vert_{\alpha=0} = (N_1,\text{ say})$
0.25	$X\vert_{\alpha=0.25} = (N_2,\text{ say})$
0.50	$X\vert_{\alpha=0.50} = (N_3,\text{ say})$
0.75	$X\vert_{\alpha=0.75} = (N_4,\text{ say})$
1	$X\vert_{\alpha=1} = (N_5,\text{ say})$

The right reference function obtained after interpolation as,

$$G(X) = \frac{(X-N_1)(X-N_3)(X-N_4)(X-N_5)0.25}{(N_2-N_1)(N_2-N_3)(N_2-N_4)(N_2-N_5)} + \frac{(X-N_1)(X-N_2)(X-N_4)(X-N_5)0.50}{(N_3-N_1)(N_3-N_2)(N_3-N_4)(N_3-N_5)}$$
$$+ \frac{(X-N_1)(X-N_2)(X-N_3)(X-N_5)0.75}{(N_4-N_1)(N_4-N_2)(N_4-N_3)(N_4-N_5)} + \frac{(X-N_1)(X-N_2)(X-N_3)(X-N_4)}{(N_5-N_1)(N_5-N_2)(N_5-N_3)(N_5-N_4)},$$

$$\frac{Qe^{-\frac{1}{2}[\frac{(ha_2+Add_2)+Br\sqrt{\pi C_p d_2(b_2-c_2)}}{a_2(a_z x^{b_z}+c_z)}]^2}}{\pi a_2 a_y x^{0.9071}(a_z x^{b_z}+c_z)} \leq X \leq \frac{Qe^{-\frac{1}{2}[\frac{(ha_3+Add_1)+Br\sqrt{\pi C_p d_1(b_1-c_3)}}{a_3(a_z x^{b_z}+c_z)}]^2}}{\pi a_1 a_y x^{0.9071}(a_z x^{b_z}+c_z)}$$

Thus the generalized membership function of $C(x,0,0)$ would be given by,

$$\mu_C(X) = \begin{cases} F(X), & \dfrac{Qe^{-\frac{1}{2}[\frac{(ha_1+Add_3)+Br\sqrt{\pi C_p d_3(b_3-c_1)}}{a_1(a_z x^{b_z}+c_z)}]^2}}{\pi a_3 a_y x^{0.9071}(a_z x^{b_z}+c_z)} \leq X \leq \dfrac{Qe^{-\frac{1}{2}[\frac{(ha_2+Add_2)+Br\sqrt{\pi C_p d_2(b_2-c_2)}}{a_2(a_z x^{b_z}+c_z)}]^2}}{\pi a_2 a_y x^{0.9071}(a_z x^{b_z}+c_z)} \\[2em] G(X), & \dfrac{Qe^{-\frac{1}{2}[\frac{(ha_2+Add_2)+Br\sqrt{\pi C_p d_2(b_2-c_2)}}{a_2(a_z x^{b_z}+c_z)}]^2}}{\pi a_2 a_y x^{0.9071}(a_z x^{b_z}+c_z)} \leq X \leq \dfrac{Qe^{-\frac{1}{2}[\frac{(ha_3+Add_1)+Br\sqrt{\pi C_p d_1(b_1-c_3)}}{a_3(a_z x^{b_z}+c_z)}]^2}}{\pi a_1 a_y x^{0.9071}(a_z x^{b_z}+c_z)} \\[2em] 0, & \text{otherwise.} \end{cases}$$

According to the *randomness-fuzziness consistency principle* [9, 10, 12], the left reference function $F(X)$ and right reference function $G(X)$ are distribution and complementary distribution functions respectively. Differentiating these distribution functions we would have the respective probability density functions, say $f(x) = \frac{d}{dx}F(X)$ and $g(x) = \frac{d}{dx}(1-G(X))$ respectively. Thus we need two probability spaces to define a possibility space.

5 Membership Functions by the Lagrange Method of Interpolation using Numerical Data

Tables 1 and 2 show the uncertain and the crisp inputs to the model $C(x,0,0)$ and the concentration is evaluated for downwind distance $x = 1500$ m away along the horizontal direction. These crisp input data are taken from the analysis of Yegnan [16], considering some degree of variability with reference to triangular fuzzy numbers.

Using the numerical data, the membership functions and curves of concentration $C(x,0,0)$ along the downwind distance for $x = 1500$ m, under unstable (Category A), neutral (Category D) and stable (Category F) categories are enumerated in the following sections.

Parameters	Lower Bound	Middle Value	Upper Bound
Wind Speed $u(m/sec)$	2.27	3.57	4.6
Discharge Velocity $w_0(m/sec)$	7.2	11.8	16.4
Ambient Temperature $T_a(K)$	273	276	279
Gas Exit Temperature $T_s(K)$	315	355	395

Table 1. Triangular fuzzy numbers representing uncertain parameters

Parameters	Values
Specific Heat $C_p(Cal/kg-K)$	0.24
Physical Stack Height $h(m)$	55
Emission Rate $Q(gm/sec)$	1
Stack Diameter $d(m)$	4

Table 2. Crisp input data

5.1   Membership function for the unstable category

The membership function and membership curve (Fig. 1) of pollutant concentration $C(x,0,0)$ for the unstable category under the Gaussian plume model are as follows

$$\mu_C(X) = \begin{cases} 4.5793 \times 10^{26}X^4 - 3.36 \times 10^{20}X^3 + 2.6 \times 10^{13}X^2 + 37100000X - 7.2097, \\ \qquad\qquad\qquad\qquad 2.309 \times 10^{-7} \leq X \leq 2.997 \times 10^{-7} \\ 7.67375 \times 10^{25}X^4 - 1.57 \times 10^{20}X^3 + 1.2767 \times 10^{14}X^2 - 51472460X + 8.5728, \\ \qquad\qquad\qquad\qquad 2.997 \times 10^{-7} \leq X \leq 4.724 \times 10^{-7} \\ 0, \qquad\qquad\qquad\qquad\qquad \text{otherwise.} \end{cases}$$

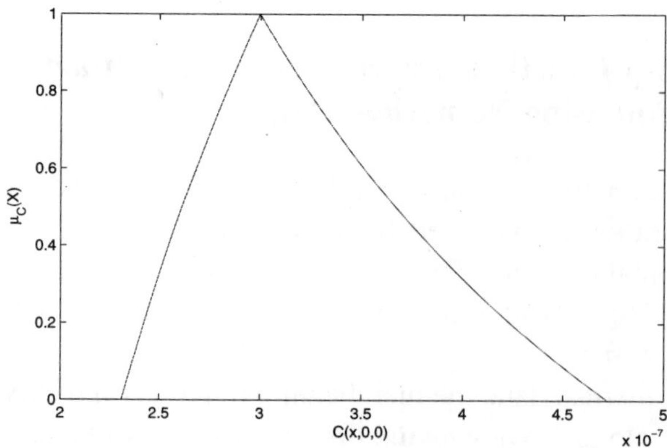

Fig. 1. Membership curve of concentration for unstable category

The left and right reference functions gives the probability density functions respectively as

$$f(x) = \begin{cases} 18.3172 \times 10^{26}x^3 - 10.08 \times 10^{20}x^2 + 5.2 \times 10^{13}x + 37100000, \\ \qquad\qquad 2.309 \times 10^{-7} \le x \le 2.997 \times 10^{-7} \\ 0, \qquad\qquad\qquad \text{otherwise,} \end{cases}$$

and

$$g(x) = \begin{cases} -30.695 \times 10^{25}x^3 + 4.716 \times 10^{20}x^2 - 2.5534 \times 10^{14}x + 51472460, \\ \qquad\qquad 2.997 \times 10^{-7} \le x \le 4.724 \times 10^{-7} \\ 0, \qquad\qquad\qquad \text{otherwise.} \end{cases}$$

5.2   Membership function for the neutral category

The membership function and membership curve (Fig. 2) of pollutant concentration $C(x,0,0)$ under the Gaussian plume model for the neutral category of weather are as follows

$$\mu_C(X) = \begin{cases} 2.82568 \times 10^{20}X^4 - 8.65 \times 10^{15}X^3 + 7.97 \times 10^{10}X^2 + 76895.69X - 0.91988, \\ \qquad\qquad 3.5163 \times 10^{-6} \le X \le 6.3240 \times 10^{-6} \\ 3.70795 \times 10^{19}X^4 - 2.34 \times 10^{15}X^3 + 6.02 \times 10^{10}X^2 - 808918.26X + 4.24225, \\ \qquad\qquad 6.3240 \times 10^{-6} \le X \le 1.13825 \times 10^{-5} \\ 0, \qquad\qquad\qquad \text{otherwise.} \end{cases}$$

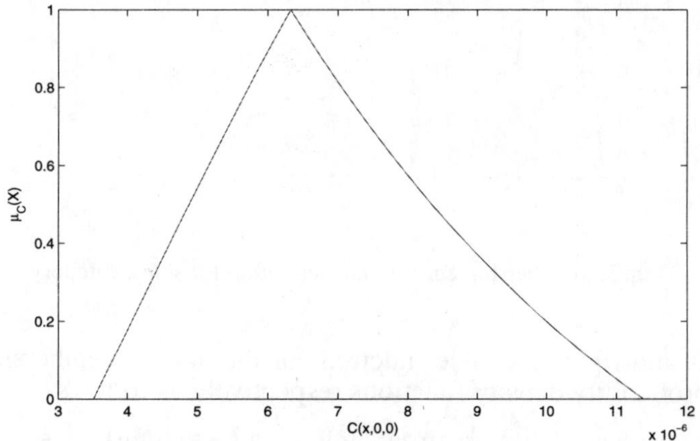

Fig. 2. Membership curve of concentration for neutral category

The left and right reference functions of the membership function gives the respective probability density functions as

$$f(x) = \begin{cases} 11.30272 \times 10^{20}x^3 - 25.95 \times 10^{15}x^2 + 15.94 \times 10^{10}x + 76895.69, \\ \qquad\qquad\qquad 3.5163 \times 10^{-6} \leq x \leq 6.3240 \times 10^{-6} \\ 0, \qquad\qquad\qquad\qquad \text{otherwise}, \end{cases}$$

and

$$g(x) = \begin{cases} -14.8318 \times 10^{19}x^3 + 7.029 \times 10^{15}x^2 - 12.048 \times 10^{10}x + 808918.26, \\ \qquad\qquad\qquad 6.3240 \times 10^{-6} \leq x \leq 1.13825 \times 10^{-5} \\ 0, \qquad\qquad\qquad\qquad \text{otherwise}. \end{cases}$$

5.3 Membership function for the stable category

The membership function and membership curve (Fig. 3) of pollutant concentration $C(x,0,0)$ for the stable category of weather under the Gaussian plume model are as follows

$$\mu_C(X) = \begin{cases} -4.223 \times 10^{21}X^4 + 0.07 \times 10^{18}X^3 - 4.559 \times 10^{11}X^2 + 1551490.733X - 1.555, \\ \qquad\qquad\qquad 1.5646 \times 10^{-6} \leq X \leq 5.6416 \times 10^{-6} \\ 2.624 \times 10^{18}X^4 - 2.778 \times 10^{14}X^3 + 1.046 \times 10^{10}X^2 - 189602.845X + 1.784, \\ \qquad\qquad\qquad 5.6416 \times 10^{-6} \leq X \leq 4.80871 \times 10^{-5} \\ 0, \qquad\qquad\qquad\qquad \text{otherwise}. \end{cases}$$

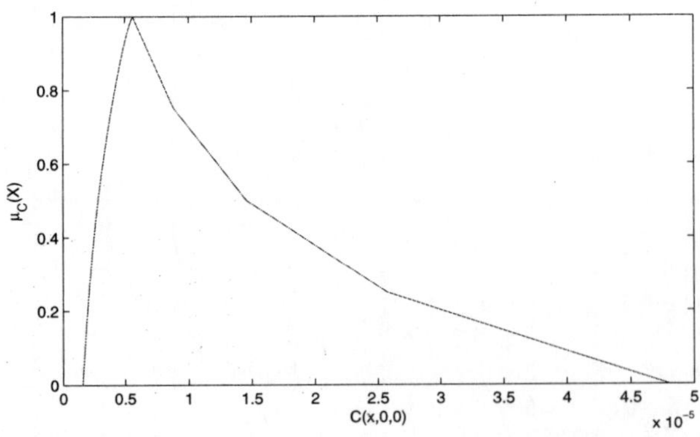

Fig. 3. Membership curve of concentration for stable category

The left and right reference functions of the above membership function yields the probability density functions respectively as follows

$$f(x) = \begin{cases} -16.8908 \times 10^{21}x^3 + 0.21039 \times 10^{18}x^2 - 9.11852 \times 10^{11}x + 1551490.733, \\ \qquad\qquad\qquad 1.5646 \times 10^{-6} \leq x \leq 5.6416 \times 10^{-6} \\ 0, \qquad\qquad\qquad\qquad \text{otherwise}, \end{cases}$$

and

$$g(x) = \begin{cases} -10.49684 \times 10^{18}x^3 + 8.33484 \times 10^{14}x^2 - 2.09264 \times 10^{10}x + 189602.8449, \\ \qquad\qquad 5.6416 \times 10^{-6} \leq x \leq 4.80871 \times 10^{-5} \\ 0, \qquad\qquad\qquad\qquad \text{otherwise.} \end{cases}$$

6 Conclusions

The membership functions of pollutant concentration of atmospheric dispersion defined by the Gaussian Plume Model can be obtained by using Lagrange's polynomial. The degree of membership function by polynomial approximation will depend on the number of discretized values of α-cuts of the concentration. The fuzziness associated with the pollutant concentration yields two independent probability laws. The left reference function is the distribution function and the right reference function is the complementary distribution function according to the *randomness-fuzziness consistency principle*, satisfying all the properties of distribution function. The probability density functions concerning the distribution functions can therefore be determined for the fuzziness associated with the pollutant concentration. Here, in our case, since we have considered five discretized values of α-cuts of the concentration, the density functions resulted in a third degree polynomial. In the literature of fuzziness, no such attempt has been made to determine the independent density functions concerning the fuzziness associated with the pollutant concentration yet.

References

1. Adel A. Abdel Rahman, On the Atmospheric Dispersion and Gaussian Plume Model, 2nd International Conference on Waste Management, Water Pollution, Air Pollution, Indoor Climate (WWAI'08), Corfu, Greece, October 26-28, 2008.
2. V.V. Shirvaikar and V.J. Daoo, Air Pollution Meteorology, Bhabha Atomic Research Centre, Mumbai, India, 2002.
3. O.G. Sutton, The problem of diffusion in the lower atmosphere, Quarterly Journal of the Royal Meteorological Society, 73(317-318) (1947), 257–281.
4. O.G. Sutton, The theoretical distribution of airborne pollution from factory chimneys, Quarterly Journal of the Royal Meteorological Society, 73(317-318) (1947), 426–436.
5. E.C. Eimuits, M.G. Konicek, Derivations of continuous functions for the lateral and vertical atmospheric dispersion coefficients, Atmospheric Environment, 16 (1972), 859–863.
6. F. Pasquill, Atmospheric Diffusion, Second Edition, Ellis Horwood Ltd., Chichester, 1961.
7. G. A. Briggs, Diffusion Estimation for small Emissions, 1973 Annual Report, Air Resources Atmospheric Turbulence and Diffusion Lab., Environmental Lab., Report ATDL-106, USDOC-NOAA, 1973.
8. F.W. Thomas, S.G. Carpenter and W.C. Colbaugh, Plume estimates for electric generating stations, Journal of Air Pollution Control Association, 20 (2) (1970), 170–177.
9. H.K. Baruah, The randomness-fuzziness consistency principle, International Journal of Energy, Information and Communications, 1(1) (2010), 37–48.

10. H.K. Baruah, The theory of fuzzy sets: beliefs and realities, International Journal of Energy, Information and Communications, 2(2) (2011), 1–22.
11. H.K. Baruah, In search of the root of fuzziness: the measure theoretic meaning of partial presence, Annals of Fuzzy Mathematics and Informatics, 2(1) (2011), 57–68.
12. H.K. Baruah, An introduction to the theory of imprecise sets: the mathematics of partial presence, Journal of Mathematics and Computer Science, 2(2) (2012), 110–124.
13. H.K. Baruah, Set superimposition and its applications to the theory of fuzzy sets, Journal of the Assam Science Society, 40(1-2) (1999), 25–31.
14. H.K. Baruah, Construction of the membership function of a fuzzy number, ICIC Express Letters, 5(2) (2011), 545–549.
15. H.K. Baruah, Construction of normal fuzzy numbers using the mathematics of partial presence, Journal of Modern Mathematics Frontier, 1(1) (2012), 9–15.
16. A. Yegnan, D.G. Williamson and A.J. Graettinger, Uncertainty analysis in air dispersion modelling, Environmental Modelling and Software, 17 (2002), 639–649.

Team Selection for Twenty20 World Cup following the Indian Premier League

Hemanta Saikia* and Dibyojyoti Bhattacharjee**

Department of Business Administration
Jawaharlal Nehru School of Management Studies
Assam University, Silchar – 788 011, Assam.

Abstract. A cricket squad generally comprises of fifteen players having different cricketing abilities, out of which XI players are fielded for a given match. The selection of XI players out of 15 is done by the team management comprising of captain, vice captain and the coach. However, selecting the squad of 15 players is done by a group of selectors (usually appointed by the concerned cricket board) based on the recent performances of a large collection of players. Selecting balanced squad, considering cricketers with different abilities like batting, bowling, wicket keeping, etc. is a very difficult decision-making problem. In 2009 and 2010 seasons of ICC Twenty20 World Cup, followed the Indian Premier League (IPL). As a large number of Indian players participated in IPL, therefore, the selectors were provided with an opportunity of selecting the optimal squad for the global event. But in both the World Cups team India had to bite the dust. This study tries to examine that if the selection of the Indian cricket team in the said seasons of ICC Twenty20 World Cup are pertinent, using a binary (0-1) integer optimization method. The study finds that the actual Indian team that was sent to ICC Twenty20 World Cup in both the years were not optimal. This optimization technique would be helpful for the optimal team selection of other team sports also.

Key words: Cricket, decision making, integer programming, optimization, performance measure, sport.

1 Introduction

Innovation is a vital aspect of the game's continued development to remain relevant and to attract new fans and supporters [1]. One suitable example of this is the new Twenty20 format of cricket and probably it is the most significant development in twenty first century. Just a few years, since the introduction of Twenty20 format of cricket in English domestic cricket, in 2007, the International Cricket Council (ICC) organized the first Twenty20 World Cup in South Africa. Surprisingly, India that had left four best players behind won the tournament. This proved to be an incredible boost to this type of cricket in India. Thereafter, Board of Control for Cricket in India (BCCI) set up a professional league so-called Indian Premier League (IPL).

---

 * Email: h.saikia456@gmail.com
** Email: djb.stat@gmail.com

In April 2008, the game of cricket got a new dimension when BCCI initiated the Indian Premier League (IPL), a Twenty20 cricket tournament to be played among eight domestic teams, named after eight Indian states or cities but owned by the franchisees [2]. The franchisees formed their teams by competitive bidding from a collection of Indian and international players and the best of Indian upcoming talent. IPL attracted the world's top players and showcased the best of India's talent. However, when team India failed to put up a valiant show in the 2009 and 2010 seasons of Twenty20 World Cup, it was the IPL, which was held responsible by the fans, critics and some veteran players of the game [3]. In the year 2009 and 2010, the national tournament IPL was followed by the Twenty20 World Cup. The cricket fans believed that experience of Indian players in the IPL would come in handy. Nevertheless, the team India's debacle performance at the said Twenty20 World Cups raised two crucial questions. First, is IPL responsible for the performances of cricketers in Twenty20 World Cups? The second is that whether team selection in the said Twenty20 World Cups were pertinent. The first issue was explicitly discussed by Saikia and Bhattacharjee [3] and found that IPL is not responsible for the performance of Indian cricketers in 2009 and 2010 ICC Twenty20 World Cup. Again, since a large number of Indian players participated in IPL, so there is a prospect for Indian selectors to select the optimal 15 players for the global event. Therefore, this paper especially discusses the second issue for the selection of an optimal Indian cricket team in 2009 and 2010 seasons of ICC Twenty20 World Cup, to examine whether the selection of the Indian cricket team is pertinent, using an integer optimization method.

2 Performance Measure

In cricket, traditional performance statistics like batting average, strike rate, bowling average, economy rate, etc. are mostly used to measure the performance of batsman and bowler. But these statistics have severe limitations in assessing the true abilities of a player's performance [4]. Combining the traditional performance statistics, several performance measures have already been developed by different authors like Lewis [4], Barr and Kantor [5], Lemmer [6–8]. However, these existing performance measures are applicable only for measuring the performances of batting, bowling or wicket keeping abilities separately. To overcome these limitations the following performance measure is proposed that can be used to measure the performance of the players based on the different abilities through scorecard of the match.

The performance measure of the ith player is given by,

$$S_i = S_{i1} + \theta_i \tag{1}$$

where $\theta_i = S_{i2}^{a_i} + S_{i3}^{1-a_i} - 1$, if i^{th} player is either a bowler or wicket keeper. Otherwise $\theta_i = 0$, and a_i is an indicator variable with $a_i = 1$ (0), if i^{th} player is a bowler (wicket keeper) and S_{i1} is the performance score for batting, S_{i2} is the performance score for bowling and S_{i3} is the performance score for wicket keeping.

2.1 Factors considered for measure the abilities

To measure the batting, bowling and wicket keeping abilities of the cricketers in Twenty20 cricket a number of factors are considered. These factors are number of innings played, batting average, strike rate and average percentage contribution to the team total for batting abilities; number of innings played, bowling average, economy rate and bowling strike rate for bowling abilities; and number of matches played, number of catches taken per match, number of stumping per match and number of bye runs conceded per match for wicket keeping abilities. For each of the abilities (i.e. batting, bowling and wicket keeping) the values of all these factors for each player are normalized and then weights of these factors are calculated based on their relative importance. All the normalized scores for considered factors are multiplied by their corresponding weights and then added together to get S_{i1}, S_{i2} and S_{i3}.

2.2 Normalization

The developed performance measure is a linear combination of traditional performance measures. However, Lewis [4] mentioned that the traditional measures of performances do not allow to combining the abilities of batting and bowling as they are based on incompatible scales. Therefore, to overcome this limitation use of normalization is essential in this model.

Let X_{ijk} be the observed values of the ith player for the jth factor of the kth ability. Out of the different factors, some are having positive dimension like batting average, batting strike rate, etc. as they are directly related to the batting ability of the player. While some of the factors like economy rate, number of bye runs conceded, etc. have negative dimension as they are negatively related to the abilities of the player. Now if the factor represents positive dimension then it is normalized as

$$Y_{ijk} = \frac{X_{ijk} - \min(X_{ijk})}{\max(X_{ijk}) - \min(X_{ijk})} \tag{2}$$

and if the factor represents negative dimension then it is normalized as

$$Y_{ijk} = \frac{\max(X_{ijk}) - X_{ijk}}{\max(X_{ijk}) - \min(X_{ijk})} \tag{3}$$

2.3   Determination of weights

The next step would be to determination of weights for different factors in each of the IPL. Iyenger and Sudarshan [9] assumed that the weights vary inversely as the variation in the respective variables. It has been applied in this study to determine the weights of different factors that are associated with the various abilities of the cricketers.

Let Y_{ijk} be the normalized value of the ith players for the jth factors of the kth ability. If w_{jk} represents the weight of the jth factor under the kth ability then it is calculated as,

$$w_{jk} = \frac{C_k}{\sqrt{var(Y_{ijk})}} \quad j = 1,2,3,4 \quad and \quad k = 1,2,3 \tag{4}$$

where $\sum_{j=1}^{4} w_{jk} = 1$ for all k and C_k is a normalizing constant that follows

$$C_k = \left(\sum_{j=1}^{4} \frac{1}{\sqrt{var(Y_{ijk})}} \right)^{-1} \tag{5}$$

The choice of the weights in this manner would ensure that the large variation in any one of the factor would not unduly dominate the contribution of the rest of the factors [9]. One can see the weights of different factors in Appendix-A.

3   Data and Computation of Performance Score

The data related to the performance of the players in the 2009 and 2010 seasons of IPL are collected from the website *www.espncricinfo.com*. To measure the performance of players it is necessary that players' statistics for large number of games should be considered because the expectations regarding level of performance can't be gauged fairly from only one match [10]. Thus, some selection criterion needs to be set up while considering the players. Only Indian players are selected based on the following criteria from the said seasons of IPL.

For batsmen, those who had played (i) at least five innings played in IPL (ii) at least 120 balls faced and (iii) had a batting average greater than 15 are considered. The bowlers who had played (i) at least five matches in IPL (ii) at least 120 balls delivered and (iii) dismissed at least 3 wickets are included. The players who had satisfied the above mentioned conditions for both the batsmen and bowlers are considered as all-rounders. For wicket keepers who had played (i) at least five matches (ii) dismissed at least 3 batsmen and (iii) satisfied all the three conditions of the batsmen are considered for the study. The 48 Indian players are selected from IPL 2009 and 52 Indian players are selected from IPL 2010 including batsmen, bowlers, all-rounders and wicket keepers. All these players are considered for the study and as per their cricketing abilities, the performance scores are computed as follows.

The batting performance ($k=1$) of the ith player is calculated by

$$S_{i1} = \sum_{j=1}^{4} w_{j1} Y_{ij1} \tag{6}$$

The bowling performance ($k=2$) of the ith player is calculated by

$$S_{i2} = \sum_{j=1}^{4} w_{j2} Y_{ij2} \tag{7}$$

The wicket keeping performance ($k=3$) of the ith player is calculated by

$$S_{i3} = \sum_{j=1}^{4} w_{j3} Y_{ij3} \tag{8}$$

On obtaining the values of S_{i1}, S_{i2} and S_{i3} the performance score S_i of the ith player is computed using equation (1). The performance scores of all the players are then converted into corresponding performance index (P_i) and for i^{th} player it is given by,

$$P_i = \frac{S_i}{\max(S_i)} \tag{9}$$

The performance index for each player is a number lying between zero and one (i.e. $0 < P_i \leq 1$). Higher values of the performance index better is the performance of the players.

4   The Optimization Model

The application of binary integer-programming model for selection of a team was proposed by Garber and Sharp [11] in order to select a limited over squad of 15 players instead of a playing XI. Extending the same idea, authors like Lourens [12], Brettenny [13], Lemmer [14], etc. used integer programming to select an optimal squad for various tournaments. In this work, the optimization technique used for team selection, in 2009 and 2010 seasons of ICC Twenty20 World Cup after the corresponding seasons of IPL is a binary (0-1) integer-programming model. The solution to the problem is attained using the Solver add-in available in Microsoft Excel.

Thus, once the values of performance scores (P_i) for the entire selected squad of Indian players are calculated then the next step is to select the optimal 15 players for the corresponding Twenty20 World Cups. However, there are some cricketing requirements that shall be followed while the different combinations are tried for optimization. Keeping in view these cricketing requirements, an optimal team is selected bestowing to various abilities of the cricketers as if two openers, at least three specialist batsmen, at least two all-rounders, at least three fast bowlers, at least two spin bowlers and one wicket keepers. A set of binary variables are defined that shall be helpful in equating the objective function and the constraints.

$\theta_i = 1(0)$, if the ith player is selected for the optimum 15 (Otherwise)
$b_i = 1(0)$, if the ith player is an opening batsman (Otherwise)
$c_i = 1(0)$, if the ith player is an specialist batsman but not opener(Otherwise)
$d_i = 1(0)$, if the ith player is a spinner (Otherwise)
$e_i = 1(0)$, if the ith player is a fast bowler (Otherwise)
$f_i = 1(0)$, if the ith player is a wicket keeper (Otherwise)
$g_i = 1(0)$, if the ith player is an all-rounder (Otherwise)

Let k be the total number of potential players who are considered for selection, then the function that is to be maximized is given by,

$$Z = \sum_{i=1}^{k} \theta_i P_i \tag{10}$$

For the current problem, the following constraints are used.

$\sum_{i=1}^{k} \theta_i = 15$ #to ensure that exactly 15 players are selected
$\sum_{i=1}^{k} \theta_i b_i = 2$ #to ensure that exactly 2 openers are selected
$\sum_{i=1}^{k} \theta_i c_i \geq 3$ #to ensure that atleast 3 specialist batsmen are selected
$\sum_{i=1}^{k} \theta_i g_i = 2$ #to ensure that exactly 2 all-rounders are selected
$\sum_{i=1}^{k} \theta_i e_i \geq 3$ #to ensure that atleast 3 fast bowlers are selected
$\sum_{i=1}^{k} \theta_i d_i \geq 2$ #to ensure that atleast 2 spin bowlers are selected
$\sum_{i=1}^{k} \theta_i f_i = 1$ #to ensure that exactly 1 wicket keeper is selected

5 Analysis and Result

In IPL 2009, out of 47 players 4 are opening batsmen, 8 specialist batsmen, 6 all-rounder, 16 fast bowler, 12 spin bowler and 4 wicket keeper. Similarly, out 51 players in IPL 2010, 7 are opening batsmen, 13 specialist batsmen, 5 all-rounder, 15 fast bowler, 12 spin bowler and 4 wicket keeper. Now the optimal 15 players for the parallel seasons of ICC Twenty20 World Cup are selected based on performances of the players in the said IPL seasons. Excel add-in 'Solver' is used to maximize the objective function Z subject to the given cricketing requirements (i.e., constraints).

Now based on results of the optimization model some players like R Dravid, S Tendulkar and A Kumble in the year 2009 and S Ganguly and S Tendulkar in the year 2010 are selected for corresponding Twenty20 World Cup. However, S Tendulkar keep himself out from this format the game of cricket since inaugural edition of Twenty20 World Cup in 2007. The legendary leg spinner A Kumble also announced his retirement from international cricket on March 3, 2007. Though V Sehwag was selected for ICC Twenty20 World Cup in 2010 but he missed it along with fast bowler P Kumar for fitness related issues. In addition, Indian selectors were not interested to select R Dravid and S Ganguly as a part of Indian squad because of their inconsistent performance in Twenty20 format of cricket. Therefore, these players should not be part of the optimum player selection process. Accordingly, we run the optimization model yet again for each of the season, to select the optimal Indian squad. The results of the selection model are given in Table 1.

Twenty20 World Cup 2009		Twenty20 World Cup 2010	
Actual Squad	*Optimum Squad*	*Actual Squad*	*Optimum Squad*
V Sehwag	V Sehwag (@)	M Vijay	M Vijay (@)
G Gambhir	G Gambhir (@)	S Raina	S Raina (*)
S Raina	S Raina (*)	Y Singh	Y Singh (*)
R Sharma	R Sharma (*)	H Singh	H Singh ($)
H Singh	H Singh ($)	Z Khan	Z Khan (#)
RP Singh	RP Singh (#)	MS Dhoni	R Uthappa (w)
P Ojha	P Ojha ($)	Y Pathan	AB Dinda (#)
Y Singh	V Kohli (+)	RA Jadeja	S Tiwary (+)
Y Pathan	A Mishra ($)	G Gambhir	S Badrinath (@)
I Shamra	A Nehra (#)	P Kumar	V Kohli (+)
P Kumar	NV Ojha (+)	A Nehra	SB Jakati (#)
MS Dhoni	KD Karthik (w)	KD Kartik	AT Rayudu (+)
RA Jadeja	P Chawla ($)	P Chawla	R Ashwin ($)
Z Khan	TL Suman (+)	R Sharma	SK Trivedi (#)
I Pathan	M Patel (#)	R Vinay Kumar	P Ojha ($)

Table 1. Optimal Indian Squad in ICC Twenty20 World Cup

(Note: w = wicket keeper, @ = opener, + = batsman, * = all-rounder, # = fast bowler and $= spin bowler)

6   Discussion

In Table 1, between actual and optimum Indian team squad, 7 and 5 players are common respectively in 2009 and 2010 seasons of ICC Twenty20 World Cup. The common players for the season 2009 are V Sehwag, G Gambhir, S Raina, R Sharma, H Singh, RP Singh and P Ojha and in 2010, the common players are M Vijay, S Raina, Y Singh, H Singh and Z Khan. Thus, it is revealed that the squad of team India sent by BCCI in the said seasons of ICC Twenty20 World

Cup were not optimal. Note that MS Dhoni is not selected as a wicket keeper in both the seasons of Twenty20 World Cup based on the given optimization method (*cf.* Table 1). However, the Indian selectors were not ready to take such decision to eliminate the person from optimal squad, who clinched the inaugural Twenty20 World Cup title in 2007 for India. Selection of captain is a vital decision for a team. In addition, the leadership qualities are difficult to quantify. As MS Dhoni, shall claim his position in the squad being the captain, so the last constraint (i.e., exactly one wicket keeper) that was used for optimization shall be removed. In the optimal selection, KD Karthik and RV Uthappa are selected as wicket keeper for the seasons 2009 and 2010 ICC Twenty20 World Cup respectively. MS Dhoni's selection as captain in the squad will restrict the inclusion of Karthik's and Uthappa's in the team, in spite of their objective selection.

7   Conclusion and Future Scope of Study

Selecting the optimal squad keeping in view the cricketing requirements, from available players having different abilities is a difficult task. Along with subjectivity, many other factors are involved in the selection of a cricket team [14]. For example - Inclusion of a player in the optimal squad of a team is depend on the strength of the opponent, pitch condition, format of the game, experience of the players, etc. Also sometimes selectors may want to give another chance to an experienced player who is out of form [14]. The difficulty of such subjective state of affairs can be traded by using an objective method that applied in this paper. However, the selection of the players in the optimal squad is subject to the set of given constraints. If the constraints are changed then players in the optimal squad may also get altered.

The optimization technique discussed in this paper helps to decide the optimal squad of a given team objectively, for any given tournament based on the previous performances of the players. The validity of the method can be tested when 2012 ICC Twenty20 World Cup becomes functional. Another area of interest may be to decide an optimum XI from a given squad of players given the team of the opposition and the pitch conditions. Finally, with the support of avowal given by Lemmer [14] that to select an optimal squad before the any given tournament, close co-operation between the selectors and the cricket statisticians are needed.

References

1. E. Mani, A strong sport growing stronger: A perspective on the growth, development and future of international cricket, Sport in Society, 12(4/5) (2009), 681–693.
2. H. Saikia, D. Bhattacharjee, On Classification of all-rounders of the Indian Premier League (IPL): A Bayesian Approach, Vikalpa, 36(4) (2011), 25–40.

3. H. Saikia, D. Bhattacharjee, Is IPL responsible for cricketers' performance in Twenty20 world cup?, International Journal of Sports Science and Engineering, 6 (2012), 96–110.
4. A.J. Lewis, Towards fairer measures of player performance in one-day cricket, Journal of the Operational Research Society, 56(7) (2005), 804–815.
5. G.D.I. Barr, B.S. Kantor, A criterion for comparing and selecting batsmen in limited overs cricket, Journal of the Operational Research Society, 55(12) (2004), 1266–1274.
6. H.H. Lemmer, The combined bowling rate as a measure of bowling performance in cricket, South African Journal for Research in Sport, Physical Education and Recreation, 24(2) (2002), 37–44.
7. H.H. Lemmer, A measure for the batting performance of cricket players, South African Journal for Research in Sport, Physical Education and Recreation, 26(1) (2004), 55–64.
8. H.H. Lemmer, Performance measure for wicket keepers in cricket, South African Journal for Research in Sport, Physical Education and Recreation, 33(3) (2011), 89–102.
9. N.S. Iyenger, P. Sudarshan, A method of classifying regions from multivariate data, Economic and Political Weekly, (1982) 2048-2052.
10. P.J. Bracewell, K. Ruggiero, A parametric control chart for monitoring individual batting performance in cricket, Journal of Quantitative Analysis in Sports, (2009) 1–19.
11. H. Garber, G.D. Sharp, Selecting a limited overs cricket squad using an integer programming model, South African Journal for Research in Sport, Physical Education and Recreation, 28(2) (2006), 81–90.
12. M. Lourens, Integer Optimization for the Selection of a Twenty20 Cricket Team, Nelson Mandela Metropolitan University, Faculty of Science, South Africa: Unpublished Master of Science Dissertation in Mathematical Statistics, 2009.
13. W. Bretteny, Integer optimization for the selection of a fantasy league cricket team, South Africa: M.Sc. Dissertation, Nelson Mandela Metropolitan University, 2010.
14. H.H. Lemmer, Team selection after a short cricket series, European Journal of Sport Science, DOI:10.1080/17461391.2011.587895, (2011) 1–7.

Appendix A

Factors for batting	IPL2009 weights	IPL2010 weights
Number of innings	0.266	0.237
Batting average	0.252	0.266
Batting strike rate	0.252	0.236
Average percentage of contribution	0.23	0.261
Factors for bowling		
Number of innings	0.222	0.217
Bowling average	0.263	0.263
Economy rate	0.243	0.228
Bowler's strike rate	0.271	0.293
Factors for wicket keeping		
Number of matches played	0.25	0.25
Number of catches taken	0.25	0.25
Number of stumping	0.25	0.25
Number of bye runs conceded	0.25	0.25

Table 2. Weights of the Different Factors in IPL

Water Wave Scattering by Small Undulation of the Porous Bottom in a Two-layer Fluid

Srikumar Panda and S. C. Martha

Department of Mathematics
Indian Institute of Technology Ropar
Nangal Road, Rupnagar – 140 001, Punjab, India.

Abstract. The problem involving scattering of water waves by small undulation of the porous sea bed in a two-layer fluid, is investigated within the framework of two-dimensional linearized water wave theory where the upper layer is free to the atmosphere. In such a two-layer fluid there exist waves with two different wave numbers (modes): one with lower wave number propagate in the upper layer whilst those with higher wave number propagate in the lower layer. An incident wave of a particular wave number gets reflected and transmitted by the bottom undulation into waves of both modes. Perturbation analysis in conjunction with Fourier transform technique is used to derive the first-order corrections of velocity potentials, reflection and transmission coefficients for both the modes due to incident waves of two different modes. One special type of bottom topography is considered as an example to evaluate the related coefficients in detail. These coefficients are depicted in graphical forms.

Key words: Water wave scattering, two-layer fluid, irrotational flow, porosity, linear theory, perturbation analysis, fourier transform, reflection and transmission coefficients.

1   Introduction

Scattering of waves in a multi-layered fluid has recently attracted to many scientists and engineers for many investigations in the areas of coastal and marine engineering. The transfer of energy from surface to internal waves was a major focus of many previous investigations.

In the absence of any obstacle, the wave propagation in a two-layer fluid with a free surface was first investigated by Stokes [1]. For a two-layer of incompressible and inviscid fluid with upper layer fluid having a free surface, Lamb [2] has shown that for a given frequency, time-harmonic small amplitude gravity waves with lower wave number propagate along the free surface while those with higher wave number propagate along the interface. When a train of progressive wave of a particular mode encounter an obstacle, it is partially reflected and partially transmitted into waves of both the modes. Thus there is a transfer of energy from surface mode to interface mode and vice versa. Linton and McIver [3] solved the problem involving water waves scattering by a long horizontal cylinder present in either of the two layers in a two-layer fluid using the method of multipoles and obtained the reflection and transmission coefficients for different wave numbers due to incident wave of different wave numbers. Then Linton and Cadby [4] extended the problem of Linton and Mciver [3] to oblique water wave scattering and solved this particular problem by multipole expansion. Maiti and Mandal [5] solved the problem involving scattering of oblique waves in a two-layer fluid by using Green's integral theorem. The motivation behind these problems arose while modelling an under water pipe bridge across one of the Norwegian fjords which consist of a layer of fresh water on top of a deep layer of salt water.

In practical coastal engineering, porosity of the bed becomes an exceedingly important aspect, as reported in the literature. Mase and Takeba [6], Zhu [7] and Silva *et al.* [8] considered water-wave reflection and/or transmission problems where a porous medium was assumed to lie on a sea bed of varying quiescent depth. Martha *et al.* [9] considered the problem of water wave scattering by small undulation on a porous bed.

In this article, we consider wave scattering by small undulation on a porous sea bed in a two-layer fluid whose upper layer is free to the atmosphere. In this case, time-harmonic progressive waves of a particular frequency can propagate with two different wave numbers: waves with lower wave number propagate in the upper layer and the other with higher wave number propagate in the lower layer. The motion of the fluid inside the porous bed is not analyzed here and it is assumed that the fluid motions are such that the bottom condition as is used here holds good and depends on a known parameter G, called porosity parameter as reported in the note by Martha *et al.* [9]. Using linear theory, the physical problem is formulated as a coupled boundary value problem (BVP) for the two potentials functions describing the fluid motion in each of the two layers. By employing perturbation analysis involving a small parameter $\varepsilon(<< 1)$ being present in the representation of the small undulation of the porous sea bed, the governing coupled BVP are reduced to a simpler BVPs upto the first order. The BVPs at the first order are solved by using Fourier transform technique to obtain the first-order velocity potentials and these potentials are utilized to obtain the first-order reflection and transmission coefficients in terms of integrals involving the shape function $c(x)$ representing the bottom undulation. From the application point of view, a patch of sinusoidal ripples is taken as an example to evaluate the integrals for reflection and transmission coefficients in detail. The numerical values for the first order reflection and transmission coefficients of both modes due to normally incident waves of two different modes are computed and depicted graphically against the wave number.

2   Mathematical Formulation of the Problem

Assuming linear theory, we consider the problem of water wave scattering by small undulation on a porous bed in a two-layer fluid in which the upper layer is free to the atmosphere. In each layer fluid is inviscid, incompressible, immiscible having a constant but different densities. The effect of surface tension at the interface of two fluids is neglected. A right-handed rectangular Cartesian co-ordinate system is considered in which x-axis is the position of the undisturbed interface and y-axis is positive vertically downward. Here the undulating porous bed is described by $y = H + \varepsilon c(x)$, where $c(x)$ is a bounded and continuous function, describing the shape of the undulation and $c(x) \to 0$ as $|x| \to \infty$ so that the sea is of uniform finite depth H far away from the undulation on either side; and $\varepsilon (<< 1)$ is a small parameter giving a measure of the smallness of the undulation. The free surface being linearized about $y = -h$. The density of the fluid in the upper layer $(-h \le y \le 0)$ is denoted by ρ_1 and the density of the fluid in the lower layer $(0 \le y \le H + \varepsilon c(x))$ is denoted by $\rho_2 > \rho_1$. Assuming the motion in each layer to be irrotational and simple harmonic in time t with angular frequency σ, it can be described by velocity potentials $Re[\psi(x,y)e^{-i\sigma t}]$ for upper layer and $Re[\phi(x,y)e^{-i\sigma t}]$ for lower layer. Within each fluid, the equation of continuity yields

$$\nabla^2 \psi = 0, \ \text{in} \ -h \le y \le 0, \quad \nabla^2 \phi = 0, \ \text{in} \ 0 \le y \le H + \varepsilon c(x), \qquad (1)$$

the linearized free surface condition is

$$K\psi + \psi_y = 0, \ \text{on} \ y = -h \ \text{with} \ K = \sigma^2/g, \qquad (2)$$

the interface conditions are

$$\psi_y = \phi_y, \ \text{on} \ y = 0, \rho(K\psi + \psi_y) = K\phi + \phi_y, \ \text{on} \ y = 0 \ \text{with} \ \rho = \rho_1/\rho_2 (< 1) (3)$$

and the bottom condition is

$$\phi_n - G\phi = 0, \quad \text{on} \ y = H + \varepsilon c(x), \qquad (4)$$

where ∇^2 is the two-dimensional Laplacian operator, g is the acceleration due to gravity, G is the porous effect parameter corresponding to the sea bed; and $\partial/\partial n$ denotes the normal derivative at a point (x,y) on the bottom. Now the relation (4) can be approximated as

$$\frac{\partial \phi}{\partial y} - \varepsilon \frac{\partial}{\partial x}\left[c(x)\frac{\partial \phi}{\partial x}\right] - G\left[\phi + \varepsilon c(x)\frac{\partial \phi}{\partial y}\right] + O(\varepsilon^2) = 0 \quad \text{on} \ y = H. \qquad (5)$$

Within this framework in two-layer fluid system, a train of time-harmonic propagating waves take the form

$$\psi(x,y) = f(k,y)e^{\pm ikx} \ \text{in} \ -h \le y \le 0, \qquad (6)$$

$$\phi(x,y) = g(k,y)e^{\pm ikx} \ \text{in} \ 0 \le y \le H, \qquad (7)$$

where

$$f(k,y) = \frac{[\sinh kH - \frac{G}{k}\sinh kH][k\cosh k(h+y) - K\sinh k(h+y)]}{K\cosh kh - k\sinh kh}, \tag{8}$$

$$g(k,y) = [\cosh k(H-y) - \frac{G}{k}\sinh k(H-y)]. \tag{9}$$

Here k is the wave number of the incident wave, which is the positive real roots of the dispersion relation

$$\Delta(k) = 0, \tag{10}$$

where

$$\Delta(k) = k^2(1-\rho) - Kk(\coth kh + \coth kH) + K^2(\rho + \coth kh \coth kH)$$

$$+2GK\coth kh\coth kH - kG(1-\rho)\coth kH - \frac{K^2 G}{k}(\coth kh + \rho\coth kH).$$

The above dispersion relation (10) has only two positive real roots m and M (say) with $m < M$, so there exist two modes of propagating waves. Wave of mode m propagates in the upper layer while the other one of mode M propagates in the lower layer.

Now we consider a normal incident wave of mode m, is of the form:

$$\psi_0(x,y) = f(m,y)e^{imx} \quad \text{in} \ -h \leq y \leq 0, \tag{11}$$

$$\phi_0(x,y) = g(m,y)e^{imx} \quad \text{in} \ 0 \leq y \leq H. \tag{12}$$

When a train of wave of mode m is incident from the direction $x = -\infty$ on the undulating bottom then the far-field behaviors of ψ and ϕ, respectively, are given by

$$\psi(x,y) \to \begin{cases} f(m,y)(e^{imx} + re^{-imx}) + Rf(M,y)e^{-iMx} & x \to -\infty, \\ \\ tf(m,y)e^{imx} + Tf(M,y)e^{iMx} & x \to \infty, \end{cases} \tag{13}$$

and

$$\phi(x,y) \to \begin{cases} g(m,y)(e^{imx} + re^{-imx}) + Rg(M,y)e^{-iMx} & x \to -\infty, \\ \\ tg(m,y)e^{imx} + Tg(M,y)e^{iMx} & x \to \infty, \end{cases} \tag{14}$$

where the unknown coefficients r, R are respectively the reflection coefficients associated with reflected waves of modes m and M, due to incident wave of mode m. Similarly, t and T denote the transmission coefficients associated with transmitted waves of modes m and M respectively, due to incident wave of mode m. These unknown coefficients are to be determined here.

3   Method of Solution

Since there are two modes of propagating wave, let us first consider a train of wave of mode m is incident to the bottom undulation.

If there is no undulation at the bottom, the incident wave train will propagate without any hindrance and there will be transmission only. In view of this along with (5), ψ, ϕ, t, r, T, R can be expressed in terms of the small parameter ε as follows:

$$\left.\begin{array}{l} \psi(x,y) = \psi_0(x,y) + \varepsilon\psi_1(x,y) + O(\varepsilon^2), \quad \phi(x,y) = \phi_0(x,y) + \varepsilon\phi_1(x,y) + O(\varepsilon^2), \\ t = 1 + \varepsilon t_1 + O(\varepsilon^2), \quad r = \varepsilon r_1 + O(\varepsilon^2), \quad T = \varepsilon T_1 + O(\varepsilon^2), \quad R = \varepsilon R_1 + O(\varepsilon^2). \end{array}\right\} \quad (15)$$

Substituting relation (15) in (1)-(3), (5), (13), (14) and then comparing the first order terms of ε on both side of the equations, we have

$$\nabla^2\psi_1 = 0, \text{ in } -h \le y \le 0, \nabla^2\phi_1 = 0, \text{ in } 0 \le y \le H, \quad (16)$$

$$K\psi_1 + \psi_{1y} = 0, \text{ on } y = -h, \quad (17)$$

$$\psi_{1y} = \phi_{1y}, \text{ on } y = 0, \quad (18)$$

$$\rho(K\psi_1 + \psi_{1y}) = K\phi_1 + \phi_{1y}, \text{ on } y = 0, \phi_{1y} - G\phi_1 = q(x), \text{ on } y = H, \quad (19)$$

$$\psi_1(x,y) \to \begin{cases} r_1 f(m,y)e^{-imx} + R_1 f(M,y)e^{-iMx} & x \to -\infty, \\ \\ t_1 f(m,y)e^{imx} + T_1 f(M,y)e^{iMx} & x \to \infty, \end{cases} \quad (20)$$

$$\phi_1(x,y) \to \begin{cases} r_1 g(m,y)e^{-imx} + R_1 g(M,y)e^{-iMx} & x \to -\infty, \\ \\ t_1 g(m,y)e^{imx} + T_1 g(M,y)e^{iMx} & x \to \infty, \end{cases} \quad (21)$$

where

$$q(x) \equiv im\frac{d}{dx}\left[c(x)e^{imx}\right] - G^2 c(x).$$

The solution of the above coupled BVP described by equations (16)-(19) is obtained by using Fourier transform technique. To solve this BVP, we decouple the BVP by replacing the condition (18) with

$$\psi_{1y} = p(x) \quad \text{on} \quad y = 0, \quad \text{and} \quad \phi_{1y} = p(x) \quad \text{on} \quad y = 0, \quad (22)$$

where $p(x)$ is an unknown function.

Now we assume that m and M to have a small positive imaginary parts so that ϕ_1 and ψ_1 decreases exponentially as $|x| \to \infty$. This ensures the existence of Fourier transforms of $\psi_1(x,y)$ and $\phi_1(x,y)$ respectively, are defined by

$$\left(\overline{\psi}_1, \overline{\phi}_1\right) = \int_{-\infty}^{\infty} \left(\psi_1, \phi_1\right) e^{-ikx} dx, \text{ with inverse } \left(\psi_1, \phi_1\right) = \frac{1}{2\pi}\int_{-\infty}^{\infty} \left(\overline{\psi}_1, \overline{\phi}_1\right) e^{ikx} dk. \quad (23)$$

92   Srikumar Panda and S. C. Martha

Applying Fourier transform and its inverse defined in relation (23) in the above BVP described in (16)-(19) and solving them, it is found that the integrands contains certain singularities and hence the residue theorem is used while employing contour integration to evaluate the first order corrections of the potentials. Comparing the first order corrections of the potentials with far-field conditions for $x \to -\infty$, the reflection coefficients are obtained as

$$r_1 = \frac{iK\{K\cosh mh - m\sinh mh\} \times (G^2 - m^2)}{\sinh mh \sinh mH \, (m\sinh mH - G\cosh mH)\Delta'(m)} \int_{-\infty}^{\infty} c(x)e^{2imx}dx \tag{24}$$

$$R_1 = \frac{iK\{K\cosh Mh - M\sinh Mh\} \times (G^2 - mM)}{\sinh Mh \sinh MH \, (M\sinh MH - G\cosh MH)\Delta'(M)} \int_{-\infty}^{\infty} c(x)e^{i(M+m)x}dx. \tag{25}$$

Similarly, when $x \to \infty$, we find out the transmission coefficients as

$$t_1 = \frac{iK\{m\sinh mh - K\cosh mh\} \times (G^2 - m^2)}{\sinh mh \sinh mH \, (m\sinh mH - G\cosh mH)\Delta'(m)} \int_{-\infty}^{\infty} c(x)dx \tag{26}$$

$$T_1 = \frac{iK\{M\sinh Mh - K\cosh Mh\} \times (G^2 - mM)}{\sinh Mh \sinh MH \, (M\sinh MH - G\cosh MH)\Delta'(M)} \int_{-\infty}^{\infty} c(x)e^{-i(M-m)x}dx. \tag{27}$$

Now when we consider a train of propagating wave of mode M to be incident on the bottom undulation the same mathematical procedure described above for the case of mode m, is followed to obtain the first-order reflection and transmission coefficients $\widehat{r}_1, \widehat{R}_1, \widehat{t}_1, \widehat{T}_1$. The final expressions of these coefficients are as follows:

$$\widehat{r}_1 = \frac{iK\{K\cosh mh - m\sinh mh\} \times (G^2 - mM)}{\sinh mh \sinh mH \, (m\sinh mH - G\cosh mH)\Delta'(m)} \int_{-\infty}^{\infty} c(x)e^{i(M+m)x}dx, \tag{28}$$

$$\widehat{R}_1 = \frac{iK\{K\cosh Mh - M\sinh Mh\} \times (G^2 - M^2)}{\sinh Mh \sinh MH \, (M\sinh MH - G\cosh MH)\Delta'(M)} \int_{-\infty}^{\infty} c(x)e^{2iMx}dx, \tag{29}$$

$$\widehat{t}_1 = \frac{iK\{m\sinh mh - K\cosh mh\} \times (G^2 - mM)}{\sinh mh \sinh mH \, (m\sinh mH - G\cosh mH)\Delta'(m)} \int_{-\infty}^{\infty} c(x)e^{i(M-m)x}dx, \tag{30}$$

$$\widehat{T}_1 = \frac{iK\{M\sinh Mh - K\cosh Mh\} \times (G^2 - M^2)}{\sinh Mh \sinh MH \, (M\sinh MH - G\cosh MH)\Delta'(M)} \int_{-\infty}^{\infty} c(x)dx. \tag{31}$$

The above integral forms can be evaluated once the bottom profile $c(x)$ is known. In the section 4, we consider a special form for the shape function $c(x)$.

In the limit of zero permeability, i.e. when the sea-bed is impermeable, the porous effect parameter (G) will vanish and the above reflection and transmission coefficients in (24)-(31) exactly match with the results of Maiti and Mandal [5].

For a two-layer fluid, there exists two energy-balance relations known as energy identities, corresponding to the incident waves of two different wave numbers. These energy-balance relations are derived by using modified Green's integral theorem. For incidence of wave with wave number m, the energy-balance relation involving r, R, t and T is

$$|r|^2 + |t|^2 + J(|R|^2 + |T|^2) = 1, \tag{32}$$

where $J = \frac{J_M}{J_m}$ with

$$J_k = ik \times \left[\rho \int_{-h}^{0} (f(k,y))^2 dy + \int_{0}^{H} (g(k,y))^2 dy \right], \quad k = m, M.$$

Similarly for the scattering of an incident wave of wave number M, the energy-balance relation can be derived as

$$|\widehat{r}|^2 + |\widehat{t}|^2 + J(|\widehat{R}|^2 + |\widehat{T}|^2) = J. \tag{33}$$

These results are useful for the checking of various numerical results involving the reflection and transmission coefficients.

4 Special Form of the Bottom Profile

Here we consider a special form for the shape function $c(x)$ in the form of a patch of sinusoidal bottom ripples as the bottom undulation. The following bottom undulation closely resembles some naturally occurring obstacles formed at the bottom due to sedimentation and ripple growth of sands:

$$c(x) = \begin{cases} a \sin \gamma x & \text{for } \dfrac{-n\pi}{\gamma} \le x \le \dfrac{n\pi}{\gamma}, \\ 0 & \text{otherwise}, \end{cases} \tag{34}$$

where n is a positive integer. This represents a patch of sinusoidal ripples with amplitude a, the patch consisting of n ripples having the same wave number γ. Since $c(x)$ is an odd function, $t_1, \widehat{T}_1$ will vanished and the other coefficients will be as follows:

$$r_1 = \frac{(-1)^n K \{K \cosh mh - m \sinh mh\} \times (G^2 - m^2)}{\sinh mh \sinh mH (m \sinh mH - G \cosh mH) \Delta'(m)} \times \frac{2a\gamma \sin \frac{2nm\pi}{\gamma}}{\gamma^2 - 4m^2}, \tag{35}$$

$$R_1 = \frac{(-1)^n K \{K \cosh Mh - M \sinh Mh\} \times (G^2 - mM)}{\sinh Mh \sinh MH (M \sinh MH - G \cosh MH) \Delta'(M)} \times \frac{2a\gamma \sin \frac{n(M+m)\pi}{\gamma}}{\gamma^2 - (M+m)^2}, \tag{36}$$

$$T_1 = \frac{(-1)^n K \{M \sinh Mh - K \cosh Mh\} \times (G^2 - mM)}{\sinh Mh \sinh MH (M \sinh MH - G \cosh MH) \Delta'(M)} \times \frac{2a\gamma \sin \frac{n(M-m)\pi}{\gamma}}{\gamma^2 - (M-m)^2}, \tag{37}$$

$$\widehat{r}_1 = \frac{(-1)^n K \{K \cosh mh - m \sinh mh\} \times (G^2 - mM)}{\sinh mh \sinh mH (m \sinh mH - G \cosh mH) \Delta'(m)} \times \frac{2a\gamma \sin \frac{n(M+m)\pi}{\gamma}}{\gamma^2 - (M+m)^2}, \tag{38}$$

$$\widehat{R}_1 = \frac{(-1)^n K \{K \cosh Mh - M \sinh Mh\} \times (G^2 - M^2)}{\sinh Mh \sinh MH (M \sinh MH - G \cosh MH) \Delta'(M)} \times \frac{2a\gamma \sin \frac{2nM\pi}{\gamma}}{\gamma^2 - 4M^2}, \tag{39}$$

$$\widehat{t}_1 = \frac{(-1)^n K \{m \sinh mh - K \cosh mh\} \times (G^2 - mM)}{\sinh mh \sinh mH (m \sinh mH - G \cosh mH) \Delta'(m)} \times \frac{2a\gamma \sin \frac{n(M-m)\pi}{\gamma}}{\gamma^2 - (M-m)^2}. \tag{40}$$

Relations (35)-(40) illustrate that for a given number of n ripples, the first order reflection and transmission coefficients due to incident wave of both modes are an oscillatory in nature. Furthermore, at the critical conditions $\gamma = 2m$ or $\gamma = M + m$ or $\gamma = M - m$ or $\gamma = 2M$, the theory predicts a resonant interaction namely Bragg resonance (see Mei [10], Alam *et al.* [11] and others), between the bed and the surface wave or interface wave.

5 Numerical Results and Discussions

The first-order reflection and transmission coefficients given in relations (35)-(40) due to incident wave of both the modes are computed numerically and are demonstrated in figures for various values of the different dimensionless parameters.

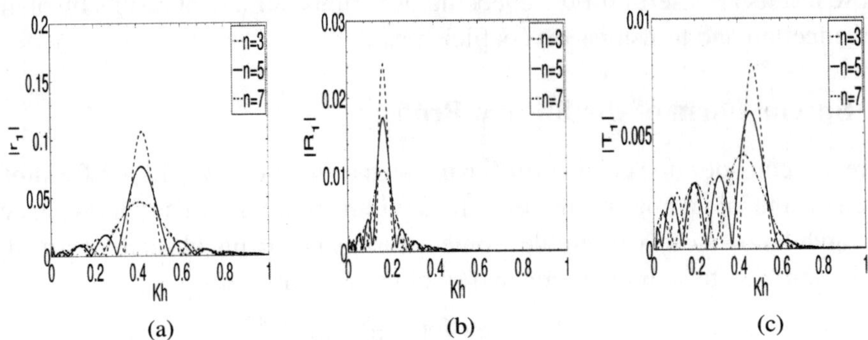

Fig. 1. (a) $|r_1|$ against Kh; (b) $|R_1|$ against Kh; (c) $|T_1|$ against Kh for $\rho = 0.5, H/h = 2, \gamma h = 1, a/h = 0.1, Gh = 0.01$.

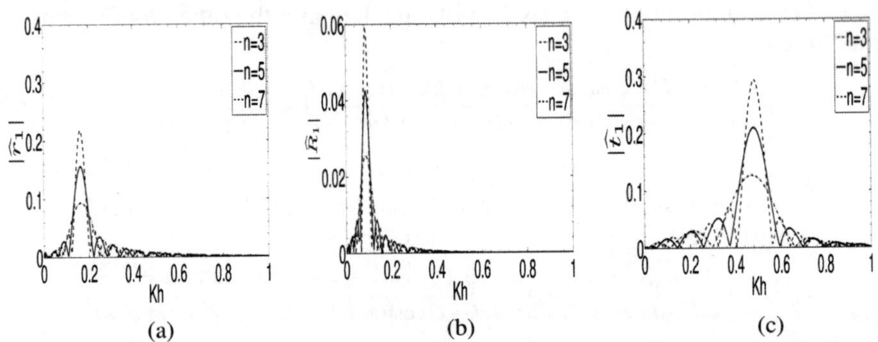

Fig. 2. (a) $|\hat{r}_1|$ against Kh; (b) $|\hat{R}_1|$ against Kh; (c) $|\hat{t}_1|$ against Kh for $\rho = 0.5, H/h = 2, \gamma h = 1, a/h = 0.1, Gh = 0.01$.

The coefficients $|r_1|, |R_1|, |T_1|, |\hat{r}_1|, |\hat{R}_1|, |\hat{t}_1|$ are shown in Figs. 1-2 against wave number Kh by considering density ratio $\rho = 0.5$, depth ratio $H/h = 2$, wave number of ripples $\gamma h = 1$, amplitude of the sinusoidal ripple $a/h = 0.1$, porous effect parameter $Gh = 0.01$ and for three different number of ripples $n = 3, 5, 7$. From these figures, it is clear that all the coefficients are oscillatory in nature as a function of wave number Kh, which validate the theoretical observation obtained in the section 4. The peak values of these coefficients are obtained respectively when $\gamma = 2m$ (the wave number of undulating bottom becomes approximately twice the surface wave number), $\gamma = (M + m)$, $\gamma = (M - m), \gamma = 2M$. In each of these figures, the peak value of the coefficients increases as the number of ripples increases. Thus, if the number of ripples increases indefinitely, the first-order coefficients become unbounded for certain value of Kh and this known as Bragg resonance, due to which the perturbation analysis fails as described by Mei [10].

The coefficients $|r_1|$, $|R_1|$, $|T_1|$, $|\widehat{r}_1|$, $|\widehat{R}_1|$, $|\widehat{t}_1|$ are plotted and shown in Figs. 3-4 against wave number Kh for three different values of porous effect parameter Gh with $\rho = 0.5, H/h = 2, \gamma h = 1, a/h = 0.1$ and $n = 3$. From these figures, it is clear that the absolute value of the coefficients at mode m (i.e. at surface wave mode) due to incident wave of both the modes decreases as the porosity increases. While the absolute value of the coefficients at mode M (i.e. at interface wave mode) due to incident wave of both the modes increases as the porosity increases.

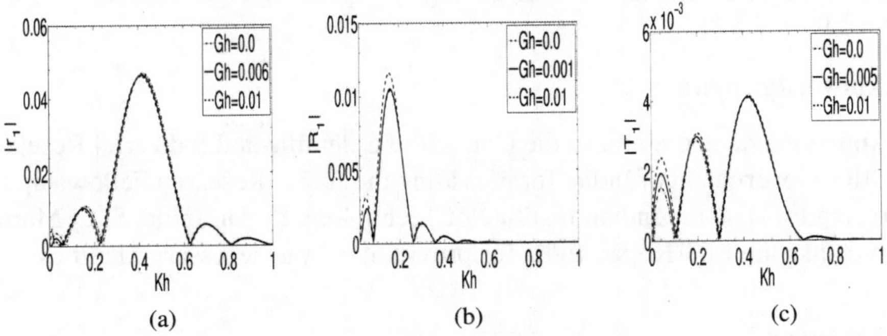

Fig. 3. (a) $|r_1|$ against Kh; (b) $|R_1|$ against Kh; (c) $|T_1|$ against Kh for $\rho = 0.5, H/h = 2, \gamma h = 1, a/h = 0.1, n = 3$.

6 Conclusion

The problem involving water wave scattering by small undulation at the porous bottom in a two-layer fluid in which the upper layer is free to the atmosphere is considered for its solution. The physical problem is formulated as coupled BVP using linear theory. Perturbation analysis in conjunction with Fourier transform technique is used to evaluate the first-order velocity potentials, reflection and transmission coefficients. Further, the numerical results for reflection and transmission coefficients are evaluated for a particular case of undulation, namely a patch of sinusoidal ripples. From the numerical results it is observed that, all the coefficients are oscillatory in nature and the peak value of these coefficients increases as the number of ripples increases. It is also clear that the absolute value of the coefficients at surface wave mode due to incident wave of both the modes decreases as the porosity increases. While the absolute value of the coefficents at interface wave mode increases as the porosity increases. The results developed here are expected to be helpful for a large class of scattering problem in a two-layer sea with an uneven porous bottom in the areas of coastal and marine engineering.

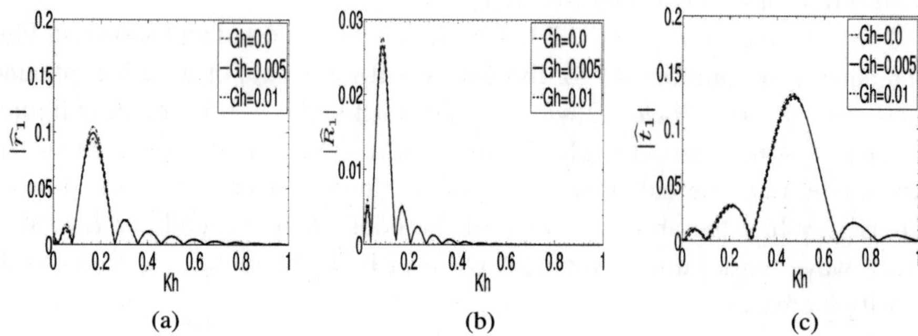

Fig. 4. (a) $|\widehat{r}_1|$ against Kh; (b) $|\widehat{R}_1|$ against Kh; (c) $|\widehat{t}_1|$ against Kh for $\rho = 0.5, H/h = 2, \gamma h = 1, a/h = 0.1, n = 3$.

Acknowledgement

Srikumar Panda is grateful to the Council of Scientific and Industrial Research (CSIR), Government of India, for providing the Junior Research Fellowship for pursuing Ph.D. at the Indian Institute of Technology Ropar, India. S. C. Martha also thanks the I.I.T Ropar, India for providing all the necessary facilities.

References

1. G.G. Stokes, On the theory of oscillatory waves, Transactions of the Cambridge Philosophical Society, 8 (1847), 441–455. (Reprinted in Mathematical Physics Papers, Cambridge University Press, 1, 314–326.)
2. H. Lamb, Hydrodynamics, 6th ed., Cambridge University Press, Cambridge, 1932.
3. C.M. Linton, M. McIver, The interaction of waves with horizontal cylinders in a two-layer fluids, J. Fluid Mech., 304 (1995), 213–229.
4. C.M. Linton, J.R. Cadby, Scattering of oblique waves in a two-layer fluid, J. Fluid Mech., 461 (2002), 343–364.
5. P. Maiti, B.N. Mandal, Scattering of oblique waves by bottom undulations in a two-layer fluid, J. Appl. Math. Comput., 22 (2006), 21–39.
6. H. Mase, K. Takeba, Bragg scattering of waves over porous ripple bed, Proc. 24th ICCE. Kobe (Japan): ASCE, 6 (1994), 635–649.
7. S. Zhu, Water waves within a porous medium on an undulating bed, Coastal Eng., 42 (2001), 87–101.
8. R. Silva, P. Salles, A. Palacio, Linear wave propagating over a rapidly varying finite porous bed, Coastal Eng., 44 (2002), 239–260.
9. S.C. Martha, S.N. Bora, A. Chakrabarti, Oblique water-wave scattering by small undulation on a porous sea-bed, Appl. Ocean Res., 29 (2007), 86–90.
10. C.C. Mei, Resonant reflection of surface waves by periodic sandbars, J. Fluid Mech., 152 (1985), 315–335.
11. M.R. Alam, Y. Liu, D.K.P. Yue, Bragg resonance of waves in a two-layer fluid propagating over bottom ripples Part I. Perturbation analysis, J. Fluid Mech., 624 (2009), 191–224.

Effect of Heat Transfer on a Steady MHD Flow over a Stretching Sheet

A. Adhikari*

Department of Mathematics,Shyampur Siddheswari Mahavidyalaya
Howrah – 711 312, West Bengal.

Abstract. The steady Magnetohydrodynamic (MHD) boundary layer viscous flow and heat transfer due to a permeable stretching sheet with prescribed surface heat flux is studied in presence of a uniform applied magnetic field. Using a special implicit finite-difference scheme, known as the Keller-box method, the nonlinear ordinary differential equations are solved. The results for the velocity and the temperature profiles are discussed for various parameters.

1 Introduction

The problem of boundary layer flow on a fixed flat plate was established by Blasius [1]. Sakiadis [2] first considered the boundary layer flow over a stretching sheet. This work was extended by Crane [3] for the two-dimensional case when the velocity is proportional to the distance from the plate. Gupta *et al.* [4] and Magyari et al. [5] studied the heat and mass transfer over a stretching sheet subject to suction or injection. With the help of integral methods, the problem of mixed convection along a vertical surface in the presence of a transverse magnetic field in a porous medium was investigated by Cheng [6]. Two-dimensional stagnation flows adjacent to a vertical heated surface with both prescribed wall temperature and prescribed wall heat flux was considered by Ramachandran *et al.* [7].

The exact solution for viscous flow induced by a shrinking sheet was investigated by Miklavcic *et al.* [8] and establish non-unique solutions for certain range of suction parameter for both two-dimensional and axisymmetric cases. Using the homotopy analysis method (HAM), Sajid *et al.* [9] extended the above problem for MHD viscous flow. Lok *et al.* [10] studied MHD stagnation point flow towards a shrinking sheet in micropolar fluid. In this paper, the steady MHD boundary layer viscous flow and heat transfer due to a permeable stretching sheet with prescribed surface heat flux is studied in presence of a uniform applied magnetic field.

2 Basic Equations

Consider a steady, three-dimensional, laminar, viscous flow of an incompressible electrically conducting fluid bounded by a stretching sheet. Here we as-

---

* Email: adhikarianimesh@gmail.com

sume that the magnetic Reynolds number is small and the electric field is zero. So the induced magnetic field is negligible. The applied magnetic field B_0 is in z-direction. The governing equations are

$$\frac{\partial u}{\partial x} + \frac{\partial v}{\partial x} + \frac{\partial w}{\partial x} = 0, \tag{1}$$

$$u\frac{\partial u}{\partial x} + v\frac{\partial u}{\partial y} + w\frac{\partial u}{\partial z} = -\frac{1}{\rho}\frac{\partial p}{\partial x} + v\left(\frac{\partial^2 u}{\partial x^2} + \frac{\partial^2 u}{\partial y^2} + \frac{\partial^2 u}{\partial z^2}\right) - \frac{\sigma B_0^2}{\rho}u, \tag{2}$$

$$u\frac{\partial v}{\partial x} + v\frac{\partial v}{\partial y} + w\frac{\partial v}{\partial z} = -\frac{1}{\rho}\frac{\partial p}{\partial y} + v\left(\frac{\partial^2 v}{\partial x^2} + \frac{\partial^2 v}{\partial y^2} + \frac{\partial^2 v}{\partial z^2}\right) - \frac{\sigma B_0^2}{\rho}v, \tag{3}$$

$$u\frac{\partial w}{\partial x} + v\frac{\partial w}{\partial y} + w\frac{\partial w}{\partial z} = -\frac{1}{\rho}\frac{\partial p}{\partial z} + v\left(\frac{\partial^2 w}{\partial x^2} + \frac{\partial^2 w}{\partial y^2} + \frac{\partial^2 w}{\partial z^2}\right), \tag{4}$$

$$u\frac{\partial T}{\partial x} + v\frac{\partial T}{\partial y} + w\frac{\partial T}{\partial z} = \alpha\left(\frac{\partial^2 T}{\partial x^2} + \frac{\partial^2 T}{\partial y^2} + \frac{\partial^2 T}{\partial z^2}\right), \tag{5}$$

The boundary Conditions are:

$$at \quad z = 0: \quad u = ax, \quad v = a(m-1)y, \quad w = -W, \quad \frac{\partial T}{\partial z} = -\frac{q_w}{\kappa}, \tag{6}$$

$$at \quad z \to \infty: \quad u = 0 = v, \quad \frac{\partial u}{\partial z} = 0 = \frac{\partial v}{\partial z}, \quad T = T_\infty. \tag{7}$$

where (u, v, w) the fluid velocity, $v = \frac{\mu}{\rho}$ the kinematic viscosity, μ the dynamic viscosity, σ the electrical conductivity, α the thermal diffusivity, κ the thermal conductivity, ρ the fluid density, p the fluid pressure, a>0 the stretching constant, q_w the surface heat flux, W the suction velocity, m=1 when the sheet stretches in x-direction (two-dimensional) and m=2 when the sheet stretches axisymmetrically, T_w the sheet temperature and T_∞ is the free stream temperature.

Apply the following Similarity transformations

$$u = axf'(\eta), \quad v = a(m-1)yf'(\eta), \quad w = -\sqrt{av}.mf(\eta),$$

$$\eta = \sqrt{\frac{a}{v}}.z, \quad \theta(\eta) = \frac{\kappa(T - T_\infty)}{q_w}\sqrt{\frac{a}{v}}, \tag{8}$$

into the equations (1) and (5).
Equation (1) is satisfied while equation (4) can be integrated as

$$\frac{p}{\rho} = v\frac{\partial w}{\partial z} - \frac{w^2}{2} + constant, \tag{9}$$

Equations (2), (3) and (5) can be reduced to

$$f''' - M^2 f' - (m-1)(f')^2 + m.f.f'' = 0, \tag{10}$$

$$\frac{\theta''}{P_r} + m.f.\theta' = 0, \tag{11}$$

where η the independent dimensionless similarity variable, θ the dimensionless temperature. The primes denotes the differentiation w.r.t. η. The $f'(\eta)$ and $\theta(\eta)$ give the velocity and temperature respectively.
The boundary conditions (6) and (7) respectively reduce to:

$$f(0) = s, \quad f'(0) = 1, \quad \theta'(0) = -1, \tag{12}$$

$$f'(\infty) = 0, \quad f''(\infty) = 0, \quad \theta(\infty) = 0, \tag{13}$$

where $P_r = \frac{\nu}{\alpha}$ is the Prandtl number, $s = \frac{W}{m\sqrt{a\nu}}$ is the suction parameter, and $M^2 = \frac{\sigma B_0^2}{\rho a}$ is the Hartmann number.

3   Numerical Solutions

The equations (10) and (11) subject to the boundary conditions (12) and (13) are solved numerically using an implicit finite-difference scheme scheme known as the Keller-box method $[11 - 13]$. The method has following four basic steps:

1. Reduce Equations (10) and (11) to a first order equation,
2. Write the difference equations using central differences,
3. Linearise the resulting algebraic equations by Newton's method and write them in Matrix-vector form,
4. Use the Block-tridiagonal elimination technique to solve the linear system.

4   Result and Discussion

Taking the step size $\Delta\eta = 0.01$ in η and within the interval $[0, \eta_\infty]$, where η_∞ is the boundary layer thickness, we run the programme in MATLAB code up to the desired level of accuracy which the difference between the input and output values of $f''(0)$ i.e., equal to 0.00001. Our numerical code is valid which is nearly same as the paper [9].

The effects of suction parameter s on velocity and temperature profiles are shown in Fig. 1 and Fig. 2. When the value of s increases the velocity profiles also increases (Fig. 1). But the temperature profiles decreases with increase of s and closer to the surface (Fig. 2).

The effects of Hartmann number M^2 on velocity and temperature profiles can be observed from Fig. 3 and Fig. 4. It is shown that the velocity profiles decrease with the increase of Hartmann number (Fig. 3). When $m = 1$ (the sheet stretches in x-direction two-dimensionally) the temperature profiles increase with the increase of M^2, but the reverse effect has been shown when $m = 2$ (the sheet stretches axisymmetrically) (Fig. 4).

The temperature profiles decreases with increase of Pr for both $m = 1$ and $m = 2$ (Fig. 5).This occurs because when Pr increases, the thermal diffusivity decreases, and it leads to the decrease of the energy transfer ability that decreases the thermal boundary layer.

Acknowledgement

The author gratefully acknowledge the financial support received from the UGC [Minor Research Project: PSW 145/11-12 (ERO) 25 Jan 12].

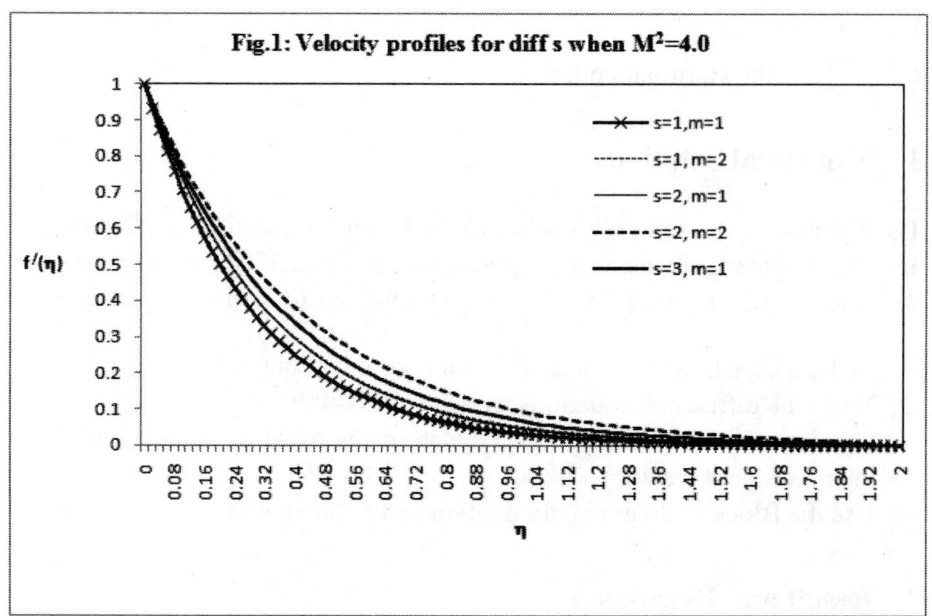

References

1. H. Blasius, Grenzschichten in Flussigkeiten mit kleiner Reibung, Zeitschrift fur Mathematik Physik, 56 (1908), 1–37.
2. B.C. Sakiadis, Boundary layer behavior on continuous solid surfaces, II. Boundary layer on a flat surfaces, AIChEJ, 7 (1961), 221–225.
3. L.J. Crane, Flow past a stretching plate, J. of Appl. Maths. And Physics (ZAMP), 21 (1970), 645–647.

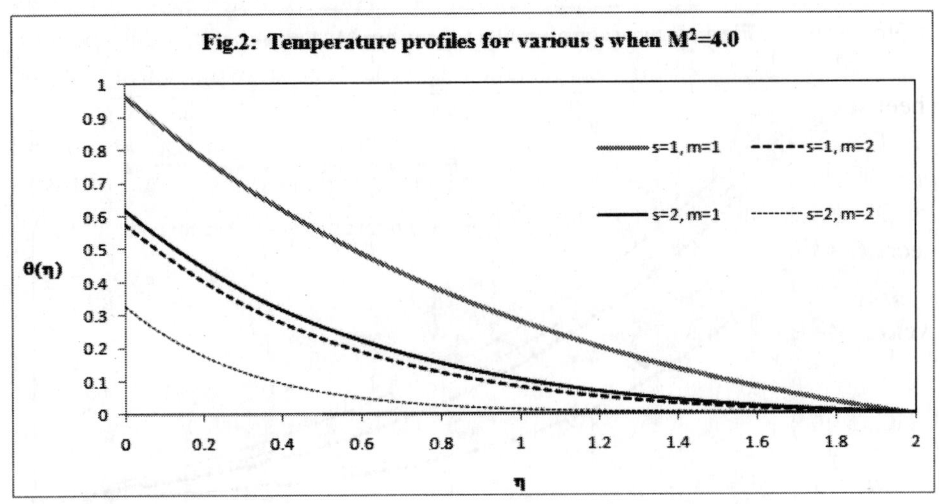

Fig.2: Temperature profiles for various s when M²=4.0
s=1, m=1
s=1, m=2
s=2, m=1
s=2, m=2
θ(η)
η

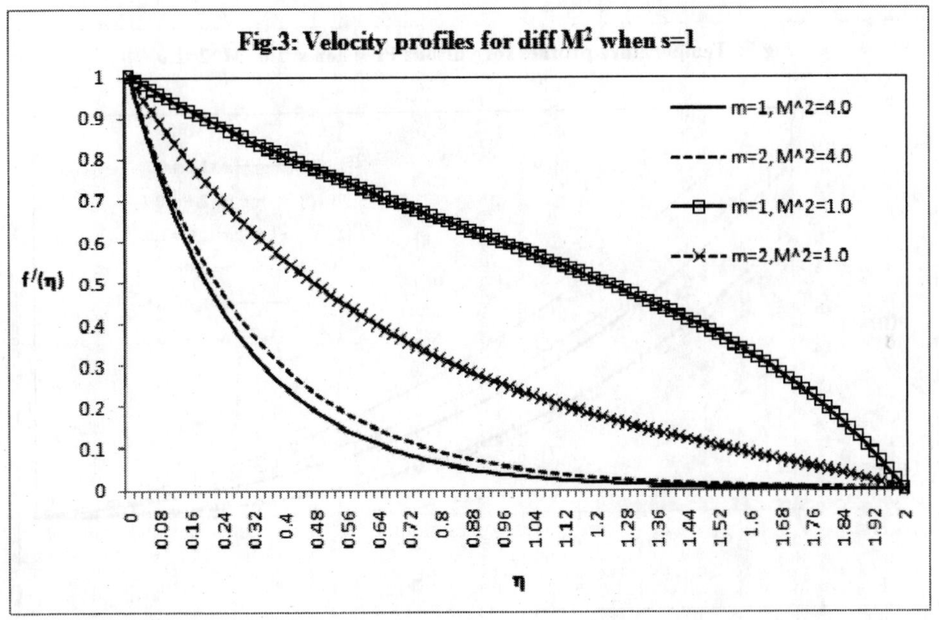

Fig.3: Velocity profiles for diff M² when s=1
m=1, M^2=4.0
m=2, M^2=4.0
m=1, M^2=1.0
m=2, M^2=1.0
f'(η)
η

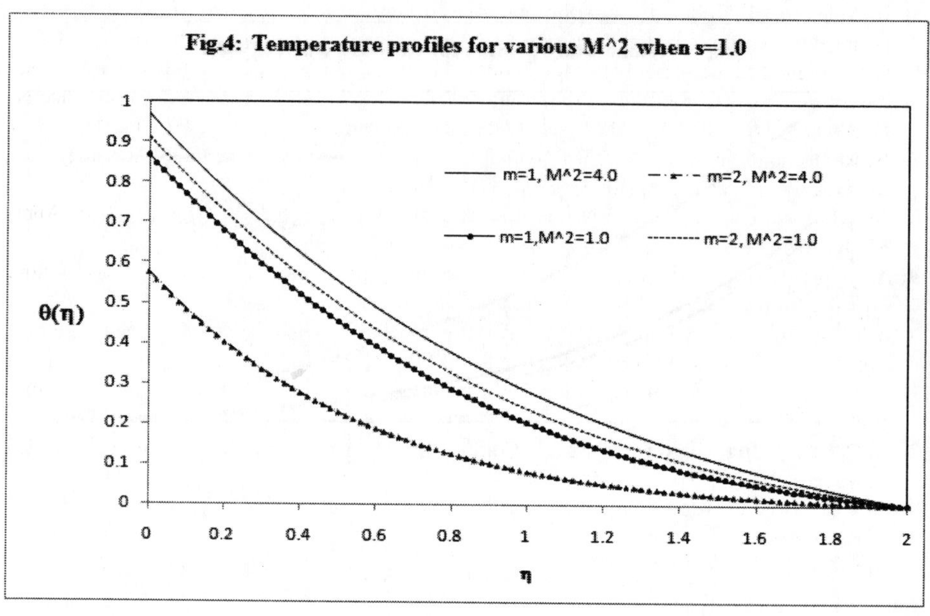

Fig.4: Temperature profiles for various M^2 when s=1.0

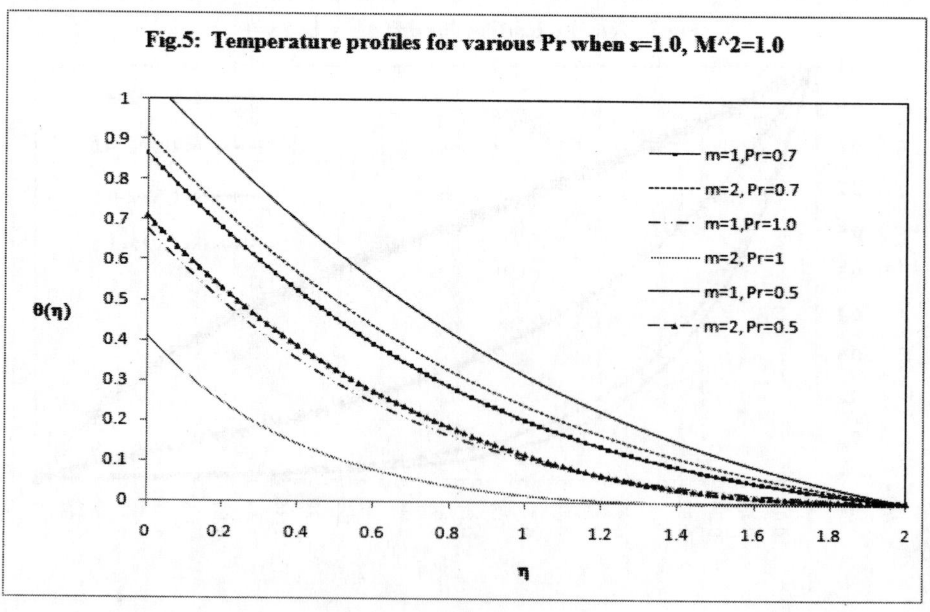

Fig.5: Temperature profiles for various Pr when s=1.0, M^2=1.0

4. P.S. Gupta, A.S. Gupta, Heat and mass transfer on a stretching sheet with suction and blowing, Canadian Jr. of Chemical Engineering, 55 (1977), 744–746.

5. E. Magyari, B. Keller, Exact solutions for self-similar boundary-layer flows induced by permeable stretching walls, European J. of Mechanics B-fluids, 19 (2000), 109–122.

6. C.Y. Cheng, Magnetic field effects on coupled heat and mass transfer embedded in a porous medium by integral methods, Proceedings of the 4th WSEAS International Conference on Heat and Mass Transfer, Gold Coast, Queensland, Australia, January 2007, 86–91.

7. N. Ramachandran, T.S. Chen, B.F. Armaly, Mixed convection in stagnation flows adjacent to vertical surfaces, J. of Heat Transfer, 110 (1988), 373–377.

8. M. Miklavcic, C.Y. Wang, Viscous flow due to a shrinking sheet, Quarterly J. of Appl. Maths., 64 (2006), 283–290.

9. M. Sajid, T. Hayat, The application of homotopy analysis method for MHD viscous flow due to a shrinking sheet, Chaos, Solitons and Fractals, 39 (2009), 1317–1323.

10. Y.Y.Lok, A. Ishak, I. Pop, MHD stagnation point flow towards a shrinking sheet, International J. for Numerical Methods in Heat Fluid Flow, 2009, Online.

11. T.Y. Na, Computational Methods in Engineering Boundary Value Problem, Academic Press, 1979.

12. T. Cebeci, P. Bradshaw, Physical and Computational Aspects of Convective Heat Transfer, Springer, 1984.

13. T. Cebeci, J. Cousteix, Modeling and Computing of Boundary-Layer Flows: Laminar, Turbulent and Transitional Boundary Layers in Incompressible and Compressible Flows, Springer, 2005.

Study of Fractional Order Two Temperature Generalized Thermo-Piezoelastic Problem

M. Islam[†,*] and M. Kanoria[‡,**]

[†]Department of Mathematics, St. Xavier's College (Evening section)
30, Mother Teresa Sarani, Kolkata – 700 016, India.
[‡] Department of Applied Mathematics, University of Calcutta
92 A.P.C. Road, Kolkata – 700 009, India.

Abstract. In this paper, a new theory of two-temperature generalized thermoelasticity is constructed in the context of a new consideration of heat conduction with fractional orders. The two-temperature Green-Naghdi (2TGNIII) model-III and two-temperature Lord-Shulman (2TLS) model of thermoelasticity are combined into a unified formulation using the unified parameters. The basic equations have been written in the form of a vector matrix differential equation in the Laplace transform domain which is then solved by using state-space approach. The inversions of Laplace transforms are computed numerically using the method of Fourier series expansion technique. The numerical estimates of the quantities of physical interest are obtained and depicted graphically. Some comparisons of the thermophysical quantities are shown in figures to estimate the effects of the fractional order parameter and the electric field.

Key words: Generalized thermopiezo-elasticity, Fractional order, vector-matrix differential equation, LS model, GN model III, Ramp-type heating.

1 Introduction

The classical theory of thermo-elasticity involving infinite speed of propagation of thermal signals, contradicts physical facts. During the last three decades, non-classical theories involving finite speed of heat transportation in elastic solids have been developed to remove this paradox. Lord and Shulman [1] incorporates a flux-rate term into Fourier's law of heat conduction and formulates a generalized form that involves a hyperbolic-type heat transport equation admitting finite speed of thermal signals. Green and Lindsay [2], which involves two relaxation times is known as temperature rate dependent thermoelasticity (TRDTE). More detailed discussions on the subject are available in the books of Ignaczak and Ostoja-starzeweski [3]. Most engineering materials such as metals possess a relatively high rate of thermal damping and thus are not suitable for use in experiments concerning second sound propagation. The relevant theoretical developments on the subject are due to Green and Naghdi [4,

---

* Email: mislam416@gmail.com
** Email: k_mri@yahoo.com

5]. Problems concerning these theory [5, 6] have been studied by many authors [7–9].

Differential equations of fractional order have been the focus of many studies due to their frequent appearance in various application in fluid mechanics, viscoelasticity, biology, physics and engineering. The most important advantage of using fractional differential equations in these and other applications is their non-local property. It is well known that the integer order differential operator is a local operator but the fractional order differential operator is non-local. This means that the next state of a system depends not only upon its current state but also upon all of its historical states. This is more realistic, and this is one reason why fractional calculus has become more and more popular [10, 11].

The first application of fractional derivatives was given by Abel who applied fractional calculus in the solution of an integral equation that arises in the formulation of the Tautochrone problem. The generalization of the concept of derivative and integral to a non-integer order has been subjected to several approaches, and some various alternative definitions of fractional derivatives appeared in [12, 13]. Recently, a considerable research effort is expended to study anomalous diffusion, which is characterized by the time-fractional diffusion-wave equation by Kimmich [14] as follows

$$\rho c = \kappa I^{\xi} c_{,ii},\qquad(1)$$

where ρ is the mass density, c the concentration, κ the diffusion conductivity, i the coordinate symbol, which takes the value 1, 2, 3. The notation I^{ξ} is the Riemann-Liouville fractional integral, introduced as a natural generalization of the well-known n-fold repeated integral $I^n f(t)$ written in a convolution-type form as in [15] which is written as follows.

$$\begin{aligned} I^n f(t) &= \frac{1}{\Gamma(n)} \int_0^t (t-\tau)^{n-1} f(\tau)d\tau, \ 0 < n \le 2, \\ &= f(t), \qquad n = 0. \end{aligned}\qquad(2)$$

where $\Gamma(n)$ is the Gamma function.

According to Kimmich [14], equation (1) describes different cases of diffusion where $0 < \xi < 1$ corredsponds to weak diffusion(subdiffusion), $\xi = 1$ corresponds to normal diffusion, $1 < \xi < 2$ corresponds to strong diffusion(superdiffusion) and $\xi = 2$ corresponds to ballistic diffusion.

Gurtin and William [16, 17] have suggested that there are no a priori grounds for assuming that the second law of thermodynamics for continuous bodies involve only a single temperature, i.e., it it is more logical to assume a second law in which the entropy contribution due to heat conduction is governed by one temperature, that of the heat supply by another. Chen and Gurtin [18] and Chen et al. [19, 20] have formulated a theory of heat conduction in deformable bodies, which depends on two distinct temperatures - the conductive temperature ϕ and the thermodynamic temperature θ. The key element that sets the two temperature thermoelasticity (2TT) apart from the classical theory of thermoelasticity (CTE) is the material parameter $a(\geq 0)$, called the temperature discrepensy [19]. Specifically, if $a = 0$, then $\phi = \theta$ and the field equations of the 2TT reduce to those of CTE. The linearized version of two temperature theory (2TT) has been studied by many authors. Warren and Chen [21] have investigated the wave propagation in the two temperature theory of thermoelasticity. Iesan [22] has established uniqueness and reciprocity theorems for the 2TT.

Due to the special characteristic, Piezoelectric material can function effectively as distributed sensors and actuators for controlling structural response. In sensor applications, mechanically or thermally induced disturbances can be determined from measurement of the induced electric potential difference, whereas in actuator applications deformation or stress can be controlled through the introduction of an appropriate electric potential difference. By integrating piezoelectric elements and advanced composite materials, the potential exists for forming high-strength, high-stiffness, light-weight structures capable of self monitoring and self controlling. The theory of thermo-piezoelectricity was first proposed by Mindlin [23]. Among the early investigations in this area, Tiersten [24] derived the differential equations and boundary conditions governing the behaviour of an electrically polarizable, finitely deformable, heat conducting medium in interaction with an electric field. Sukla and Kanoria [25] have analyzed the two temperature generalized thermo-piezoelastic problem using the state-space approach.

In this paper, we have studied the variation of stress, strain and conductive temperature in a fractional order thermo-piezo-elastic half-space body under thermal loading for 2TGNIII and 2TLS models. The governing equations of the problem are solved in the Laplace transform domain by using state-space approach. The inversion of the transformed solutions are carried out numerically applying Fourier series expansion technique [26]. The results obtained theoretically have been computed numerically and are presented graphically. Comparison of the results for different thermoelastic theories are made.

2 Basic Equations and Constitutive Relations

We now consider a thin semi-infinite piezoelectric body with stress free boundary occupying the space $x \geq 0$. At the near end of the body, a thermal effect is given which raises the temperature of this end to a prescribed temperature with known function, the direction of piezoelectric being parallel to x-axis.

For one-dimensional problem we assume displacement component of the form

$$u_x = u(x,t), \; u_y = 0, \; u_z = 0. \tag{3}$$

Then the strain components in our case become

$$e = \frac{\partial u}{\partial x}.\tag{4}$$

In the context of linear theory of Generalized thermoelasticity in absence of body forces and heat sources, the constitutive equation, the equation of motion, the Gauss's law and electric field relation and the heat equation in unified form, can be written as

$$\sigma = c_{11}\frac{\partial u}{\partial x} - \beta\theta - hD,\tag{5}$$

$$c_{11}\frac{\partial^2 u}{\partial x^2} - \beta\frac{\partial\theta}{\partial x} = \rho\frac{\partial^2 u}{\partial t^2},\tag{6}$$

Since there is no free charge inside the piezoelectric body from Gauss law, we have

$$\frac{\partial D}{\partial x} = 0. \text{ Hence } D = \text{constant},\tag{7}$$

$$E = -\frac{\partial\varphi}{\partial x},\tag{8}$$

$$I^{\xi-1}K\frac{\partial^3\phi}{\partial x^2\partial t} + I^{\xi-1}K^*\frac{\partial^2\phi}{\partial x^2} = \rho c_E\left(\frac{\partial^2\theta}{\partial t^2} + t_1\tau_0\frac{\partial^3\theta}{\partial t^3}\right) + \beta\theta_0\left(\frac{\partial^3 u}{\partial t^2\partial x} + t_1\tau_0\frac{\partial^4 u}{\partial t^3\partial x}\right)\tag{9}$$

The conductive temperature ϕ is given by

$$\phi - \theta = a\frac{\partial^2\phi}{\partial x^2}\tag{10}$$

where ϕ the conductive temperature, θ the thermodynamic temperature, a the temperature discrepancy, D the electric displacement component, θ_0 the reference temperature, ρ the density, c_E the specific heat, β the stress temperature coefficient, K^* an additional material constant, K the thermal conductivity, τ_0 the thermal relaxation. For $t_1 = 1, K^* = 0$ the equation (9) reduces to LS model and for $t_1 = 0$ the equation (9) reduces to GNIII model.

Also $0 < \xi < 1$ for weak conductivity, $\xi = 1$ for normal conductivity, and $1 < \xi \leq 2$ for strong conductivity.

We introduce the following dimensionless quantities:

$$u' = c_0\eta u,\ x' = c_0\eta x,\ t' = c_0^2\eta t,\ \sigma' = \frac{\sigma}{c_{11}}, \theta' = \frac{\theta}{\theta_0},\ \tau_0' = c_0^2\eta\tau_0,$$

$$\phi' = \frac{\phi}{\theta_0},\ D' = \frac{h}{c_{11}}D,\ t_0' = c_0^2\eta t_0,\ K^{*'} = \frac{K^*}{\rho c_E c_0^2},\ c_0^2 = \frac{c_{11}}{\rho},\ \eta = \frac{\rho c_E}{K}.$$

Omitting primes, Eqs. (5), (6), (9) and (10) can be re-written in dimensionless form as

$$\sigma = e - \alpha\theta - D,\tag{11}$$

$$\frac{\partial^2 e}{\partial x^2} - \alpha \frac{\partial^2 \theta}{\partial x^2} = \frac{\partial^2 e}{\partial t^2}, \tag{12}$$

$$\left[I^{\xi-1} K^* + I^{\xi-1} \frac{\partial}{\partial t} \right] \nabla^2 \phi = \left(1 + t_1 \tau_0 \frac{\partial}{\partial t} \right) \left[\frac{\partial^2 \theta}{\partial t^2} + \varepsilon \frac{\partial^2}{\partial t^2} \left(\frac{\partial u}{\partial x} \right) \right], \tag{13}$$

$$\theta = \phi - \omega \frac{\partial^2 \phi}{\partial x^2}, \tag{14}$$

where $\alpha = \dfrac{\beta \theta_0}{c_{11}}$, $\varepsilon = \dfrac{\beta}{\rho c_E}$, $\omega = a c_0^2 \eta^2$.

The half-space with quiescent initial state is subjected to ramp-type heating in the boundary plane $x = 0$ in the form

$$\left. \begin{aligned} F(t) &= 0, \ t \leq 0 \\ &= \frac{F_0}{t_0} t, \ 0 < t \leq t_0 \\ &= F_0, \ t > t_0 \end{aligned} \right\}. \tag{15}$$

where F_0 is constant and t_0 the ramp type parameter.

The bounding plane $x = 0$ traction free, which gives

$$\sigma(x,t) = 0 \quad \text{at } x = 0; \tag{16}$$

Since the medium initially is at rest and undisturbed, all the state functions are assumed to be zero.

3   Method of Solution

Applying Laplace transform to the equations (11)-(16) and writing the resulting equations in the form of vector-matrix differential equation in unknowns $\bar{\phi}$ and $\bar{e}$, we get

$$\frac{d^2 \bar{V}(x,s)}{dx^2} = A(s)\bar{V}(x,s) \tag{17}$$

where $\bar{V}(x,s) = \begin{pmatrix} \bar{\phi} \\ \bar{e} \end{pmatrix}$ and $A(s) = \begin{pmatrix} l & l\varepsilon \\ M & N \end{pmatrix}$.

$$M = \frac{\alpha(1-l\omega)l}{1 + l\varepsilon\alpha\omega}, \quad N = \frac{s^2 + \alpha(1-l\omega)l\varepsilon}{1 + l\varepsilon\alpha\omega}, \quad l = \frac{a_3}{1 + a_3\omega} \quad \text{and}$$

$$a_3 = \frac{s^{\xi+1}(1 + t_1\tau_q s + t_2\tau_q^2 s^2)}{[K^* + (1+\tau_v)s + \tau_T s^2]}.$$

Also

$$\bar{\theta} = (1 - l\omega)\bar{\phi} - l\omega\varepsilon\bar{e}. \tag{18}$$

110   M. Islam and M. Kanoria

4   State-Space Approach

The solution of equation (17) can be written in the form

$$\bar{V}(x,s) = \bar{V}(0,s)e^{-\sqrt{A(s)}\,x} \tag{19}$$

where $\bar{V}(0,s) = \begin{pmatrix} \bar{\phi}_0 \\ \bar{e}_0 \end{pmatrix}$ and

$$\left.\begin{aligned}
\bar{\phi}_0 &= \bar{F}(s) = \frac{F_0(1 - e^{-st_0})}{t_0 s^2} \\
\bar{e}_0 &= \frac{1}{1 + l\omega\varepsilon\alpha}\left[\frac{\alpha(1 - l\omega)F_0(1 - e^{-st_0})}{t_0 s^2} + \frac{D}{s}\right]
\end{aligned}\right\} . \tag{20}$$

Now using state-space approach, equation (19) can be written as

$$\bar{V}(x,s) = \Big[h_{ij}(x,s)\Big]\bar{V}(0,s), \tag{21}$$

where

$$h_{11} = \frac{(P_1^2 - l)e^{-P_2 x} - (P_2^2 - l)e^{-P_1 x}}{P_1^2 - P_2^2}, h_{12} = \frac{\varepsilon l(e^{-P_1 x} - e^{-P_2 x})}{P_1^2 - P_2^2},$$

$$h_{21} = \frac{M(e^{-P_1 x} - e^{-P_2 x})}{P_1^2 - P_2^2}, h_{22} = \frac{(P_1^2 - N)e^{-P_2 x} - (P_2^2 - N)e^{-P_1 x}}{P_1^2 - P_2^2}.$$

and P_1, P_2 are the characteristic roots of the matrix $\sqrt{A(s)}$ where,

$$\sqrt{A(s)} = \frac{1}{P_1 + P_2}\begin{pmatrix} l + P_1 P_2 & l\varepsilon \\ M_2 & M_1 + P_1 P_2 \end{pmatrix}$$

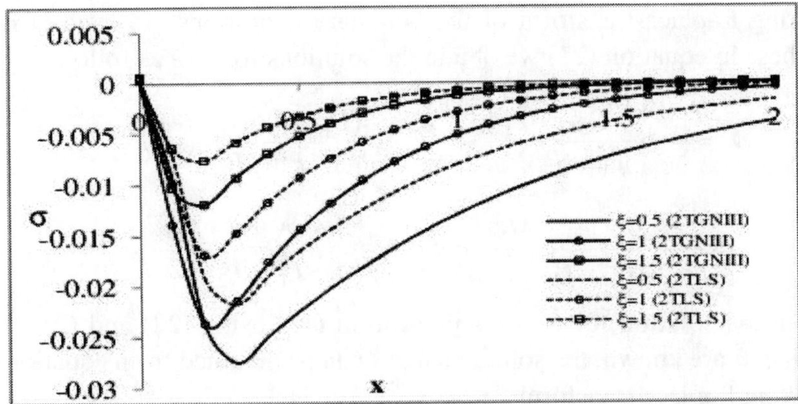

Fig.1. Variation of stress distribution σ against x for $t_0=0.20$ and $\omega=0.1$

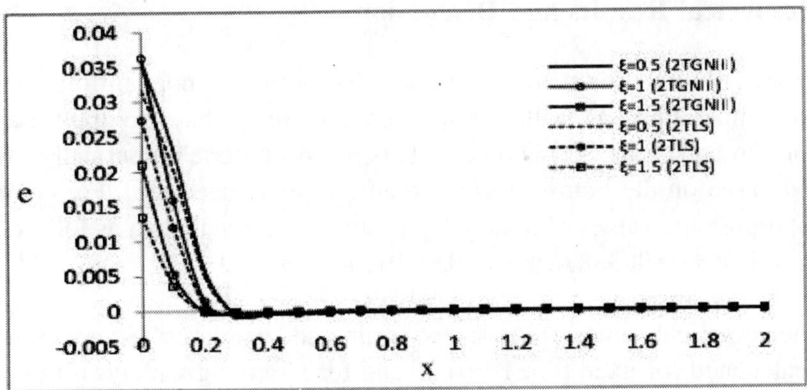

Fig.2. Variation of strain distribution e against x for $t_0=0.20$ and $\omega=0.1$

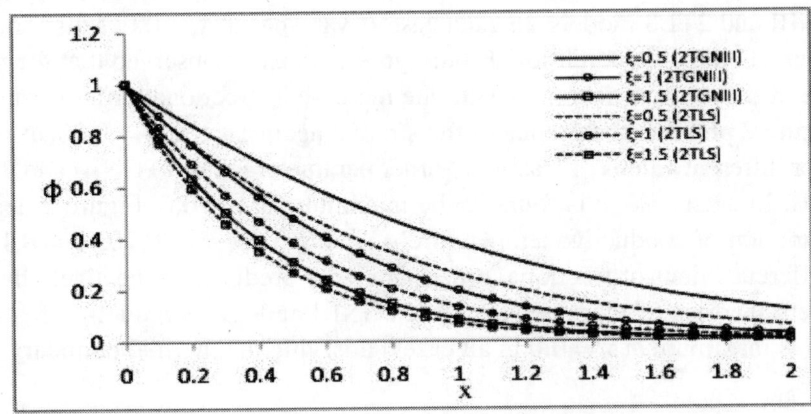

Fig.3. Variation of conductive temperature ϕ against x for $t_0=0.20$ and $\omega=0.1$

Taking Laplace-transform of the boundary conditions (15) and (16) and using these in equation (21) we obtain the solutions for $\bar{\phi}$, $\bar{e}$ as follows

$$\bar{\phi} = \frac{[(P_1^2 - l)\bar{\phi}_0 - \varepsilon l \bar{e}_0]}{P_1^2 - P_2^2} e^{-P_2 x} - \frac{[(P_2^2 - l)\bar{\phi}_0 - \varepsilon l \bar{e}_0]}{P_1^2 - P_2^2} e^{-P_1 x}, \tag{22}$$

$$\bar{e} = \frac{[(P_1^2 - N)\bar{e}_0 - M\bar{\phi}_0]}{P_1^2 - P_2^2} e^{-P_2 x} - \frac{[(P_2^2 - N)\bar{e}_0 - M\bar{\phi}_0]}{P_1^2 - P_2^2} e^{-P_1 x}. \tag{23}$$

Then the solution for $\bar{\theta}$ is obtained from (18) using (22) and (23). Now, since $\bar{\phi}, \bar{e}, \bar{\theta}$ are known, the solutions for $\bar{\sigma}$ can be obtained from equation (11) after taking Laplace transform.

5   Numerical Results and Discussion

To get the solutions for stress component (σ), strain component (e) and conductive temperature (ϕ) in the space time domain we have to apply Laplace inversion to equations (22) and (23). This has been done numerically using a method based on the Fourier-series expansion techniques [26]. For computational purpose the values of material constants have been taken as follows [27]
$F_0 = 1, \varepsilon = 0.003887, \alpha = 0.036991, \omega = 0.1, D = 10^{-7}, K^* = 7, \tau_0 = 0.01$.

The numerical values of the stress, strain and conductive temperature have been calculated for fixed time $t = 0.25$ and for x ranging widely from $x = 0.0$ up to $x = 2.0$. Figure 1 represents the variation of stress wave σ against x for $t_0 = 0.20, \omega = 0.1$ and for different values of fractional order parameter using 2TGNIII and 2TLS models. In each case σ vanishes at $x = 0.0$, satisfying the theoretical boundary condition. From figure 1 it can be observed that the magnitude of stress wave decreases with the increase of fractional order parameter ξ. Figure 2 presents the graphs of the strain e against x for $t_0 = 0.20, \omega = 0.1$ and for different values of fractional order parameter using 2TGNIII and 2TLS models. In each case, e is found to be maximum at $x = 0.0$. Figure 3 depicts the variation of conductive temperature ϕ against x for $t_0 = 0.20, \omega = 0.1$ and for different values of fractional order parameter predicted respectively by two different theories of thermoelasticity (2TGNIII and 2TLS models). It is seen that ϕ is maximum at $x = 0.0$ in all cases satisfying the thermal boundary conditions.

Acknowledgement

We are grateful to Prof. S.C. Bose of the Department of Applied Mathematics, University of Calcutta for his valuable suggestions and guidance in preparation of the paper.

References

1. H.W. Lord, Y.H. Shulman, A generalized dynamical theory of thermoelasticity, J. Mech. Phys. Solids, 15 (1967), 299–309.
2. A.E. Green, K.A. Lindsay, Thermoelasticity, J. Elasticity, 2 (1972), 1–7.
3. J. Ignaczak, M. Ostoja-Starzewski, Thermoelasticity with finite wave speeds, Oxford Science Publications, pp. 413, 2010.
4. A.E. Green, P.M. Naghdi, A re-examination of the basic results of thermomechanics, Proc. Roy. Soc. London. Ser. A, 432(1991), 171–194.
5. A.E. Green, A.E. Nagdhi, Thermoelasticity without energy dissipation, J. Elasticity, 31 (1993), 189–208.
6. A.E. Green, P.M. Naghdi, On undamped heat waves in an elastic solid, J. Therm. Stress., 15 (1992), 253–264.
7. S.H. Mallik, M. Kanoria, A two dimensional Problem for a transversely isotropic generalized thermoelastic thick plate with spatially varying heat source, Eur. J. Mech. A/Solids, 27 (2008), 607–621.
8. A. Kar, M. Kanoria, Thermo-elastic interaction with energy dissipation in an unbounded body with a spherical hole, Int. J. Solids Struct., 44 (2007), 2961–2971.
9. M. Islam, S.H. Mallik and M. Kanoria, Dynamic response in two-dimensional transversely isotropic thick plate with spatially varying heat sources and body forces, J. Appl. Math. Mech. Engl. Ed., 32 (2011), 1315–1332.
10. M. Caputo, Linear models of dissipation whose Q is almost frequently independent II. Geophys., J. R. Astron. Soc., 13 (1967), 529–539.
11. I. Podlubny, Fractional Differential Equations, Academic Press, New York, 1999.
12. R. Gorenflo, F. Mainardi, Fractional Calculus: Integral and Differential Equations of Fractional Orders, Fractals and Fractional Calculus in Continuum Mechanics, Springer, Wien 1997.
13. R. Hilfer, Application of Fraction Calculus in Physics, World Scientific, Singapore, 2000.
14. R. Kimmich, Strange Kinetics, porous media, and NMR, J. Chem. Phys., 284 (2002), 243–285.
15. F. Mainardi, R. Gorenflo, On Mittag-Lettler-type function in fractional evolution processes, J. Comput. Appl. Math., 118 (2000), 283–299.
16. M. E. Gurtin, W.O. Williams, On the clausius-duhem inequality, Z. angew Math. Phys., 7 (1966), 626–633.
17. M. E. Gurtin, W.O. Williams, An axiomatic foundation for continuum thermodynamics, Arch. Ration. Mech. Anal., 26 (1967), 83–117.
18. P. J. Chen, M. E. Gurtin, On a theory of heat conduction involving two temperatures, Z. angew Math. Phys., 19 (1968), 614–627 .
19. P. J. Chen, M. E. Gurtin and W. O. Williams, A note on non simple heat conduction, Z. angew Math. Phys., 19 (1968), 969–970.
20. P. J. Chen, M. E. Gurtin and W. O. Williams, On the thermodynamics of non-simple elastic materials with two temperatures, Z. angew Math. Phys., 20 (1969), 107–112.
21. W. E. Warren, P. J. Chen, Wave propagation in two temperatures theory of thermoelasticity, Acta Mech., 16 (1973), 83–117.
22. D. Iesan, On the linear Coupled Thermoelasticity with Two temperatures, J. Appl. Math. Phys., 21 (1970), 583–591.
23. R. D. Mindlin, On the equations of motion of piezoelectric crystals, in: N. I. Muskilishivili, Problems of continuum Mechanics, 70th Birthday Volume, SIAM, Philadelphia, 282–290, 1961.
24. H. F. Tiersten, On the nonlinear equations of thermoelectroelasticity, Int. J. Engrg. Sci., 9 (1971), 587–604.
25. S. Banik, M. Kanoria, Study of two temperature generalized thermo-piezoelastic problem, J. Thermal Stress, (2012) (In Press).

26. G. Honig, U. Hirdes, A method for the numerical inversion of Laplace transform, J. Comp. Appl. Math., 10 (1984), 113–132.

27. H. M. Youssef, E. Bassiouny, Two temperature generalized thermopiezoelasticity for one dimensional problems- State space approach, Computational Methods in Science and Technology, 14 (2008), 55–64.

Two Temperature Hyperbolic Heat Equation with Variable Thermal Conductivity

Sudip Mondal[†,*], Sadek Hossain Mallik[‡,**], M. Kanoria[§,***]

[†] Bhatkunda High School, Dist. - Burdwan, 713 153, West Bengal.
[‡] Department of Mathematics, Aliah University, Kolkata – 700 091.
[§] Department of Applied Mathematics, University of Calcutta, Kolkata – 700 009.

Abstract. A mathematical model of two-temperature generalized thermoelasticity is constructed for isotropic elastic material with variable thermal conductivity with the dual-phase-lag heat conduction law. The state space approach is adopted for the solution of one-dimensional application for an elastic half space which is subjected to mechanical and thermal loading on its boundary. The Laplace transform technique is used. The general solution obtained is applied to a specific problem when the boundary of the body of the body is subjected to thermal shock and the mechanical loading which prevents the cubical dilatation. A numerical method based on Fourier series expansion technique is employed for the inversion of the Laplace transform. Two-temperature Lord-Shulman theory can be obtained as a particular case. Numerical results for thermodynamic temperature and conductive temperature are presented by means of graphs for both the theories and are discussed.

Key words: Two temperature generalized thermoelasticity, Dual-Phase-Lag Model, variable thermal conductivity, state-space-approach.

1 Introduction

It is a well known fact that physical properties of engineering materials vary considerably with temperature. In practice there are innumerable heat conduction problems in which the temperature differences are high compared to the absolute temperatures and the thermal properties of the solid vary considerably in the operational temperature range. For this reason, the assumption of constant thermal properties sometimes leads to erroneous results during the computation. Hence, if more dependable solutions to thermal stress problems are needed, the temperature dependency of material properties should be taken into account.

Chen and Gurtin [1] and Chen et al. [2, 3] have formulated a theory of heat conduction in deformable bodies, which depends on two distinct temperatures, the conductive temperature ϕ and the thermodynamic temperature θ. The linearized version of two temperature theory (2TT) has been studied by many authors [4–7].

* Email: sudipmondal555@gmail.com
** Email: sadek_mallik@yahoo.co.in
*** Corresponding author. Email: k_mri@yahoo.com

In order to overcome the paradox of an infinite speed of thermal wave inherent in CTE and CCTE (classical coupled theory of thermoelasticity), efforts were made to modify coupled thermoelasticity, on different grounds, to obtain a wave-type heat conduction equation by different researchers, viz, Lord and Shulman [8], Green and Lindsay [9], Green and Naghdi [10, 11]. Chandrasekharaiah [12] extended the dual phase-lag model proposed by Tzou [13] to a hyperbolic thermoelastic model with dual phase-lag effects.

Youssef [14] has developed theory of two temperature generalized thermoelasticity based on LS model. Several investigations have been made by Magaňie and Quintanilla [15], Kumar et al. [16], Youssef and El Bary [17], Youssef [18], Banik and Kanoria [19, 20] based on two temperature generalized thermoelasticity theory. Several works have been done by Quintanilla [21, 22] and Quintanilla and Racke [23] considering dual-phase-lag model.

The aim of the present paper is to study the effect of variable thermal conductivity on thermoelastic interactions in a semi-infinite isotropic medium in the context of two temperature generalized dual-phase-lag thermoelasticity theory. The boundary of the medium is subjected to a thermal loading (thermal shock) and mechanical loading which is enough to prevent the cubical dilatation. The governing equations of the problem are solved in Laplace transform domain by applying state-space approach. The inversion of the transformed solutions are carried out numerically by applying a method based on Fourier series expansion technique [24] and are presented in the form of graphs for conductive temperature and thermodynamic temperature.

2   Basic Formulation

The constitutive equations are

$$\sigma_{ij} = 2\mu e_{ij} + (\lambda e - \gamma\theta)\delta_{ij}, \ i,j = 1,2,3, \tag{1}$$

where $e_{ij} = \frac{1}{2}(u_{i,j} + u_{j,i})$ and $e = e_{kk}$.

In the context of two temperature generalized thermoelasticity based on dual-phase-lag model, the equation of motion in the absence of body forces and the heat conduction equation in absence of heat sources for a linearly isotropic generalized thermoelastic solid are respectively given by

$$\rho \ddot{u}_i = (\lambda + \mu)u_{j,ji} + \mu u_{i,jj} - \gamma\theta_{,i}, \ i,j = 1,2,3, \tag{2}$$

$$\left[K\left(1 + \tau_T\frac{\partial}{\partial t} + \frac{\tau_T^2}{2}\frac{\partial^2}{\partial t^2}\right)\phi_{,i}\right]_{,i} = \left(1 + \tau_q\frac{\partial}{\partial t} + \frac{\tau_q^2}{2}\frac{\partial^2}{\partial t^2}\right)\left(\frac{K}{\kappa}\dot{\theta} + \gamma T_0\dot{e}\right), \tag{3}$$

where ρ is the density, λ and μ are Lame's constant, K is thermal conductivity, $\gamma = (3\lambda + 2\mu)\alpha_t$; α_t being the co-efficient of linear thermal expansion, T_0 is the reference temperature, $\kappa = \frac{K}{\rho c_E}$; c_E being the specific heat at constant strain, τ_q is the phase lag of heat flux vector; interpreted as the relaxation time due to fast transient effects of thermal inertia, and τ_T is the phase-lag of the temperature gradient; interpreted as the delay time caused by the microstructural interactions.

The relation between conductive temperature (ϕ) and thermodynamic temperature (θ) is given by

$$\phi - \theta = a\phi_{,ii}, \quad i = 1,2,3, \tag{4}$$

where $a(\geq 0)$ is the two temperature parameter, called temperature discrepancy.

For $\tau_T = 0$ above theory reduces to two-temperature Lord-Shulman theory [14].

We shall consider here the thermal conductivity as a linear function of thermodynamical temperature in the following form [18]

$$K(\theta) = K_0[1 + K_1\theta], \tag{5}$$

where K_0 is a constant which is equal to the thermal conductivity of the material when it does not depend on thermodynamical temperature (θ) and K_1 is a non-positive small parameter.

Now using the following mappings [18]

$$\tilde{\phi} = \frac{1}{K_0} \int_0^{\phi} K(\xi)d\xi, \tag{6}$$

$$\tilde{\theta} = \frac{1}{K_0} \int_0^{\theta} K(\xi)d\xi \tag{7}$$

Eqs. (2)-(4) can be written as follows

$$\rho \ddot{u}_i = (\lambda + \mu)u_{j,ji} + \mu u_{i,jj} - \gamma\tilde{\theta}_{,i}, \tag{8}$$

$$\left(1 + \tau_T\frac{\partial}{\partial t} + \frac{\tau_T^2}{2}\frac{\partial^2}{\partial t^2}\right)\tilde{\phi}_{,ii} = \left(1 + \tau_q\frac{\partial}{\partial t} + \frac{\tau_q^2}{2}\frac{\partial^2}{\partial t^2}\right)\left(\frac{\partial}{\partial t}\right)\left(\frac{\tilde{\theta}}{\kappa} + \frac{\gamma T_0}{K_0}e\right), \tag{9}$$

$$\tilde{\phi} - \tilde{\theta} = a\tilde{\phi}_{,ii}, \quad i = 1,2,3. \tag{10}$$

3 Formulation of the Problem

We consider a half space $(0 \leq x < \infty)$ with x axis pointing into the medium. This half space is subjected to thermal and mechanical loads on the bounding plane $(x = 0)$, that depends on the time t and are linearly quiescent. We shall consider one dimensional thermoelastic deformation of the body so that the displacement components can be taken in the following form

$$(u_x, u_y, u_z) = (u(x,t), 0, 0). \tag{11}$$

The strain displacement relation is

$$e_{xx} = \frac{\partial u}{\partial x} \tag{12}$$

and the constitutive relation (1) takes the form

$$\sigma_{xx} = (\lambda + 2\mu)\frac{\partial u}{\partial x} - \gamma\theta. \tag{13}$$

The equation of motion, heat transport equation, and relation between conductive temperature and thermodynamic temperature can be written as

$$\rho\ddot{u} = (\lambda + 2\mu)\frac{\partial^2 u}{\partial x^2} - \gamma\frac{\partial\tilde{\theta}}{\partial x}, \tag{14}$$

$$\left(1 + \tau_T\frac{\partial}{\partial t} + \frac{\tau_T^2}{2}\frac{\partial^2}{\partial t^2}\right)\frac{\partial^2\tilde{\phi}}{\partial x^2} = \left(1 + \tau_q\frac{\partial}{\partial t} + \frac{\tau_q^2}{2}\frac{\partial^2}{\partial t^2}\right)\left(\frac{\dot{\tilde{\theta}}}{\kappa} + \frac{\gamma T_0}{K_0}\frac{\partial\dot{u}}{\partial x}\right), \tag{15}$$

$$\tilde{\phi} - \tilde{\theta} = a\frac{\partial^2\tilde{\phi}}{\partial x^2}. \tag{16}$$

We now use the following non dimensional variables, to make the above equations non-dimensional.

$$x' = \frac{c_0}{\kappa}x, \; u' = \frac{c_0}{\kappa}u, \; t' = \frac{c_0^2}{\kappa}t, \; \tau_T' = \frac{c_0^2}{\kappa}\tau_T, \; \tau_q' = \frac{c_0^2}{\kappa}\tau_q, \; \sigma_{xx}' = \frac{\sigma_{xx}}{\lambda + 2\mu},$$

$$\phi' = \frac{\phi}{T_0}, \; \tilde{\phi}' = \frac{\tilde{\phi}}{T_0}, \; \theta' = \frac{\theta}{T_0}, \; \tilde{\theta}' = \frac{\tilde{\theta}}{T_0}, \; \text{where } c_0^2 = \frac{\lambda + 2\mu}{\rho}.$$

Then the corresponding non-dimensional equations, after omitting the primes, are

$$\sigma_{xx} = \frac{\partial u}{\partial x} - \varepsilon_2\theta, \tag{17}$$

$$\ddot{u} = \frac{\partial^2 u}{\partial x^2} - \varepsilon_2\frac{\partial\tilde{\theta}}{\partial x}, \tag{18}$$

$$\left(1 + \tau_T\frac{\partial}{\partial t} + \frac{\tau_T^2}{2}\frac{\partial^2}{\partial t^2}\right)\frac{\partial^2\tilde{\phi}}{\partial x^2} = \left(1 + \tau_q\frac{\partial}{\partial t} + \frac{\tau_q^2}{2}\frac{\partial^2}{\partial t^2}\right)\left(\dot{\tilde{\theta}} + \varepsilon_1\dot{e}\right), \tag{19}$$

$$\tilde{\phi} - \tilde{\theta} = \beta\frac{\partial^2\tilde{\phi}}{\partial x^2}, \tag{20}$$

where

$$\varepsilon_1 = \frac{\gamma\kappa}{K_0}, \; \varepsilon_2 = \frac{\gamma T_0}{(\lambda + 2\mu)}, \; \beta = \frac{ac_0^2}{\kappa^2}.$$

Initial and regularity conditions for the problem are given by

$$u = \theta = \phi = 0 \text{ at } t = 0 \text{ for } x \geq 0,$$

$$\frac{\partial u}{\partial t} = \frac{\partial \theta}{\partial t} = \frac{\partial \phi}{\partial t} = 0 \text{ at } t = 0 \text{ for } x \geq 0,$$

$$u = \theta = \phi = 0 \text{ as } x \to \infty.$$

Boundary conditions for the problem have been taken in the following form

$$\phi(0,t) = \phi_1 H(t),$$

where ϕ_1 is constant, and

$$e(0,t) = 0.$$

4   Method of Solution

Applying the Laplace transform defined by

$$\bar{f}(s) = \int_0^\infty e^{-st} f(t)dt, \ Re(s) > 0$$

to the equations (17)-(20), the problem can be reduced in solving the following equations

$$\frac{d^2\bar{\bar{\phi}}}{dx^2} = \alpha_1\bar{\bar{\phi}} + \varepsilon_1\alpha_1\bar{e}, \tag{21}$$

$$\frac{d^2\bar{e}}{dx^2} = \alpha_2\bar{\bar{\phi}} + \alpha_3\bar{e}, \tag{22}$$

where

$$a = \left(\frac{1 + \tau_q s + \frac{1}{2}\tau_q^2 s^2}{1 + \tau_T s + \frac{1}{2}\tau_T^2 s^2} \right)$$

$$\alpha_1 = \frac{a}{1 + a\beta}, \quad \alpha_2 = \frac{\alpha_1\varepsilon_2(1 - \beta\alpha_1)}{[1 + \beta\alpha_1\varepsilon_1\varepsilon_2]}, \quad \alpha_3 = \frac{s^2 + \alpha_1\varepsilon_1\varepsilon_2(1 - \beta\alpha_1)}{[1 + \beta\alpha_1\varepsilon_1\varepsilon_2]}.$$

5   State Space Approach

Equations (21) and (22) can be written in the form of a vector matrix differential equations as follows [25]

$$\frac{d^2\bar{V}(x,s)}{dx^2} = A(s)\bar{V}(x,s) \tag{23}$$

where

$$\bar{V}(x,s) = \begin{bmatrix} \bar{\bar{\phi}}(x,s) \\ \bar{e}(s,x) \end{bmatrix} \text{ and } A(s) = \begin{bmatrix} \alpha_1 & \varepsilon_1\alpha_1 \\ \alpha_2 & \alpha_3 \end{bmatrix}.$$

The solution of the equation (23) bounded at infinity can be obtained by applying State-space approach [25] as follows

$$\bar{\phi} = \frac{l}{s(k_1 - k_2)}[(\alpha_1 - k_2)e^{-\sqrt{k_1}x} - (\alpha_1 - k_1)e^{-\sqrt{k_2}x}] = \bar{\phi}_0 \text{ (say)}, \ K_1 = 0,$$

$$= \frac{1}{K_1}\left[\sqrt{1 + 2K_1\bar{\phi}_0} - 1\right], \ K_1 < 0,$$

$$(24)$$

$$\bar{e} = \frac{l\alpha_2}{s(k_1 - k_2)}[e^{-\sqrt{k_1}x} - e^{-\sqrt{k_2}x}], \tag{25}$$

and then the solutions for displacement component, thermodynamic temperature and stress component can be obtained as follows

$$\bar{u} = \frac{-l\alpha_2}{s(k_1 - k_2)}\left[\frac{e^{-\sqrt{k_1}x}}{\sqrt{k_1}} - \frac{e^{-\sqrt{k_2}x}}{\sqrt{k_2}}\right], \tag{26}$$

$$\bar{\theta} = \frac{l}{s(k_1 - k_2)}[(\alpha_1 - k_2)(1 - \beta k_1)e^{-\sqrt{k_1}x}$$
$$- (\alpha_1 - k_1)(1 - \beta k_2)e^{-\sqrt{k_2}x}] = \bar{\theta}_0(say), \ K_1 = 0 \tag{27}$$

$$= \frac{1}{K_1}\left[\sqrt{1 + 2K_1\bar{\theta}_0} - 1\right], K_1 < 0$$

$$\bar{\sigma}_{xx} = \frac{l\alpha_2}{s(k_1 - k_2)}[e^{-\sqrt{k_1}x} - e^{-\sqrt{k_2}x}] - \varepsilon_2\bar{\theta}. \tag{28}$$

where

$$l = \phi_1 + \frac{K_1}{2s}\phi_1^2,$$

$$\bar{\phi}_0 = \frac{l}{s(k_1 - k_2)}[(\alpha_1 - k_2)e^{-\sqrt{k_1}x} - (\alpha_1 - k_1)e^{-\sqrt{k_2}x}],$$

$$\bar{\theta}_0 = \frac{l}{s(k_1 - k_2)}[(\alpha_1 - k_2)(1 - \beta k_1)e^{-\sqrt{k_1}x} - (\alpha_1 - k_1)(1 - \beta k_2)e^{-\sqrt{k_2}x}]$$

and k_1, k_2 are the roots of the equation

$$k^2 - k(\alpha_1 + \alpha_3) + (\alpha_1\alpha_3 - \varepsilon_1\alpha_1\alpha_2) = 0.$$

This completes the solution of the problem in Laplace transform domain.

6   Numerical Results and Discussion

To get the solution for dilatation, thermal displacement, conductive temperature, thermodynamic temperature and thermal stress in the space time domain we have applied Laplace inversion formula to the equations (24), (25), (26), (27) and (28) respectively, which have been done numerically using a method based on Fourier series expansion technique [24]. The numerical code has been prepared using Fortran 77 programming language. For computational purpose copper material has been taken into consideration. The values of the material constants are taken as follows [26]

$$\lambda = 7.76 \times 10^{10} \text{Nm}^{-2}, \mu = 3.86 \times 10^{10} \text{Nm}^{-2}, \rho = 8954 \text{Kg m}^{-3},$$

$$\text{K}_0 = 386 \text{Wm}^{-1}\text{K}^{-1}, c_E = 383.1 \text{JKg}^{-1}\text{K}^{-1}, T_0 = 293K,$$

$$\alpha_t = 1.78 \times 10^{-5}K^{-1}, \varepsilon_1 = 1.618, \varepsilon_2 = 0.01041, \beta = 0.1.$$

Also for computational purpose we have taken $t = 0.2$, $\phi_0 = 1$ and $\phi_1 = 1$.

Figs. 1 and 2 are drawn to make a comparison of the results between two temperature theory (2TT) ($\beta = 0.1$) and one temperature theory (1TT) ($\beta = 0$) of generalized thermoelasticity as well as to compare between Lord-Shulman (LS) model and Dual-phase-lag (DPL) model. These two figures show variation of conductive temperature ϕ and thermodynamic temperature θ respectively against x for $K_1 = -0.5$.

Fig.1 shows that the magnitude of conductive temperature (ϕ) in one temperature ($\beta = 0$) differs slightly from two temperature theory ($\beta = 0.1$) for DPL model whereas for LS model conductive temperature ϕ is greater for one temperature theory ($\beta = 0$) than two temperature theory($\beta = 0.1$) for the values of x from 0 to 0.5. It is also seen that the magnitude of ϕ for DPL model is greater compare to the LS model for both one and two temperature theory.

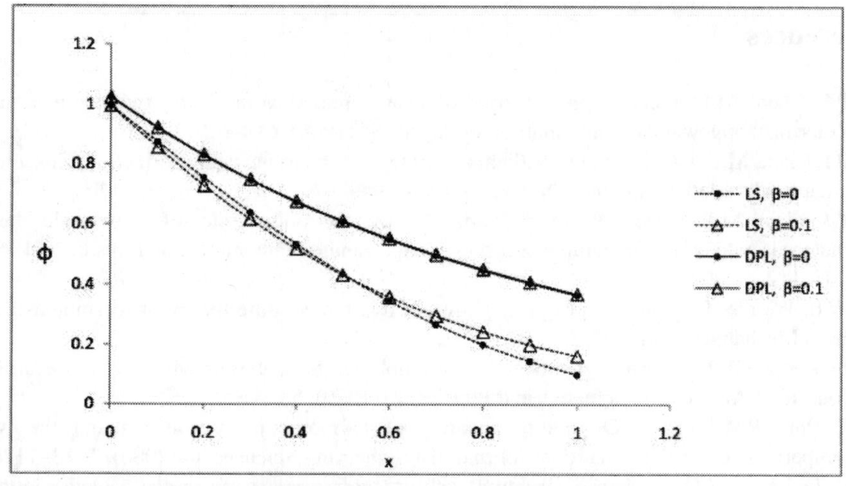

Fig.1: The conductive temperature (ϕ) distribution against distance x for K_1=-0.5

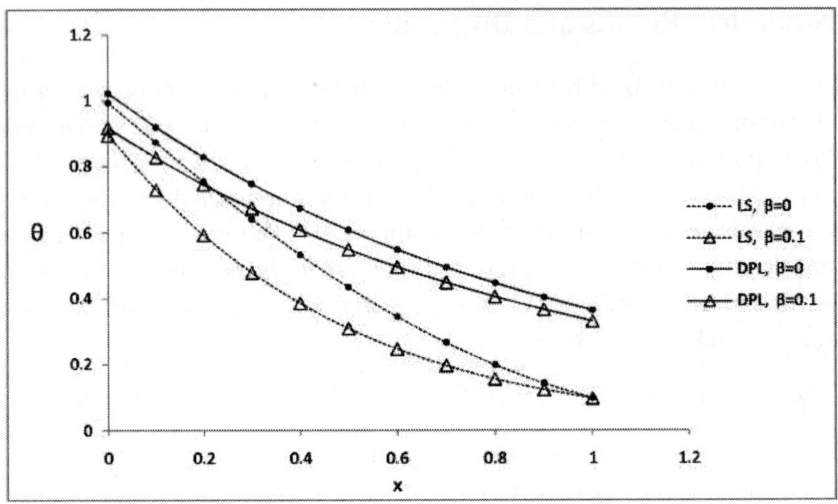

Fig.2: The thermodynamic temperature (θ) distribution against distance x for K_1=-0.5

Fig.2 shows that the values of thermodynamic temperature θ is greater for one temperature theory ($\beta = 0$) compare to two temperature theory ($\beta = 0.1$) for both DPL and LS models. It is also seen from figure that for both one temperature theory ($\beta = 0$) and two temperature theory ($\beta = 0.1$) the values of θ for DPL model is greater than that of LS model.

Acknowledgement

We are grateful to Prof. S.C. Bose of the Department of Applied Mathematics, University of Calcutta for his valuable suggestions and guidance in preparation of the paper.

References

1. P.J. Chen, M.E. Gurtin, On a theory of heat conduction involving two temperatures. Zeitschrift angewandte Mathematik und Physik, 19(1968), 614–627.
2. P.J. Chen, M.E. Gurtin, W. O. Williams, A note on non simple heat conduction. Zeitschrift angewandte Mathematik und Physik, 19(1968), 969–970.
3. P.J. Chen, M.E. Gurtin, W. O. Williams, On the thermodynamics of non-simple elastic materials with two temperatures, Zeitschrift angewandte Mathematik und Physik, 20(1969), 107–112.
4. W.E. Warren, P.J. Chen, Wave propagation in two temperature theory of thermoelasticity. Acta Mechanica, 16(1973), 83–117.
5. D. Lesan, On the thermodynamics of non-simple elastic materials with two temperatures, Journal of Applied Mathematics and Physics, 21(1970), 583–591.
6. P. Puri, P.M. Jordan, On the propagation of harmonic plane waves under the two-temperature theory, International Journal of Engineering Sciences, 44(2006), 1113–1126.
7. R. Quintanilla, On existence, structural stability, convergence and spatial behavior in thermoelasticity with two temperatures, Acta Mechanica, 168(2004), 61–73.

8. H. Lord, Y. Shulman, A Generalized dynamical theory of thermoelasticity. Journal of Mechanics and Physics of Solids, 15(1967), 299–309.

9. A.E. Green, K.A. Lindsay, Thermoelasticity. Journal of Elasticity, 2(1972), 1–7.

10. A.E. Green, P.M. Naghdi, On undamped heat waves in an elastic solid, Journal of Thermal Stresses, 15(1992), 252–264.

11. A.E. Green, P.M. Naghdi, Thermoelasticity without energy dissipation. Journal of Elasticity, 31(1993), 189–208.

12. D.S. Chandrasekharaiah, Hyperbolic thermoelasticity: A review of recent literature, Applied Mechanics Review, 51(1998), 705–729.

13. D.Y. Tzou, A unified field approach for heat conduction from macro to micro scales, ASME Journal of Heat Transfer, 117(1995), 8–16.

14. H.M. Youssef, Theory of two-temperature generalized thermoelasticity, IMA Journal of Applied Mathematics, 71(2006), 1–8.

15. A. Magaйie, R. Quintanilla, Uniqueness and growth of solutions in two-temperature generalized thermoelastic theories, Journal of Mathematics and Mechanics of Solids, 14(2009), 622–634.

16. R. Kumar, R. Prasad, S. Mukhopadhyay, Variational and reciprocal principles in two-temperature generalized thermoelasticity, Journal of Thermal Stresses, 33(2010), 161–171.

17. H.M. Youssef, A.A. El-Bary, Two temperature generalized thermoelasticity with variable thermal conductivity, Journal of Thermal Stresses, 33(2010), 187–201.

18. H.M. Youssef, State-space approach on generalized thermoelasticity for an infinite material with a spherical cavity and variable thermal conductivity subjectd to ramp-type heating, The Canadian Applied Mathematics Quarterly, Applied Mathematics Institute, 13(2005), 369–390.

19. S. Banik, M. Kanoria, Two-temperature generalized thermoelastic interactions in an infinite body with a spherical cavity, International Journal of Thermophysics, 32(2011), 1247–1270.

20. S. Banik, M. Kanoria, Effects of the three-phase-lag on two temperature generalized thermoelasticity for an infinite medium with a spherical cavity, Applied Mathematics and Mechanics -English Edition, 33(2012), 483–498.

21. R. Quintanilla, Exponential stability in the dual-phase-lag heat conduction theory, J. Non-Equilib. Thermodyn., 27(2002), 217-227.

22. R. Quintanilla, A condition on the delay parameters in the onedimensional dual-phase-lag thermoelastic theory, J. Thermal Stresses, 26(2003), 713-721.

23. R. Quintanilla, R. Racke, A note on stability in dual-phase-lag heat conduction, International Journal of Heat and Mass Transfer, 49(2006), 1209–1213.

24. G. Honig, U. Hirdes, A method for the numerical inversion of Laplace transform, Journal of Computational and Applied Mathematics, 10(1984), 113–132.

25. L.Y. Bahar, R.B. Hetnarski, State space approach to thermoelasticity, Journal of Thermal Stresses, 1(1978), 135–145.

26. H.M. Youssef, E.A. Al-Lehaibi, State-space approach of two temperature generalized thermoelasticity of one dimensional problem, International Journal of Solids and Structures, 44(2007), 1550–1562.

Multilayer Perceptron Models Based on PCA for Total Ozone Concentration Time Series

Goutami Chattopadhyay[†,*], Parthasarathi Chakraborthy[‡,**] and Surajit Chattopadhyay[‡,***]

[†]Bengal Engineering and Science University, Shibpur,
Howrah – 711 103, West Bengal.
[‡]Pailan College of Management and Technology, Bengal Pailan Park,
Kolkata – 700 104, West Bengal.

Abstract. The present paper reports a study on daily total ozone concentration time series over four metro cities of India in the multivariate environment. Introducing rotated component matrix for the principal components, the predictors suitable for generating artificial neural network (ANN) for daily total ozone prediction are identified. ANNs in the form of multilayer perceptron (MLP) trained through backpropagation learning are generated for all of the study zones and the model outcomes are assessed statistically.

1 Introduction

Neural networks, or more precisely artificial neural networks (ANN), are a branch of artificial intelligence. Multilayer perceptrons (MLP) form one type of ANN. For a systematic theoretical discussion on ANN see references [1–3]. The capability of MLP for approximating continuous functions with arbitrary accuracy has been demonstrated through several results [4–6]. The advent of the backpropagation (BP) algorithm [7], an adaptation of the steepest descent method, opened avenues for the application of multilayer neural networks for many problems of practical interest [6]. In this algorithm, an initial weight (or parameter) vector w_0 of a feedforward neural network is iteratively adapted according to the recursion [6]

$$w_{k+1} = w_k + \eta d_k \tag{1}$$

to find an optimal weight vector. This adaptation is performed by presenting to the network sequentially a set pairs of input and target vectors. The positive constant of h, which is selected by the user, is called the learning rate. The direction vector d_k is given by [6]

$$d_k = -\nabla E(w_k) \tag{2}$$

* Email: goutami15@yahoo.co.in
** Email: pschakra@yahoo.com
*** Email: surajit_2008@yahoo.co.in, surajcha@iucaa.ernet.in

Benefits that ANN offer when compared to more traditional statistical modelling techniques are well-documented in literature (e. g. [8–10] and references therein). The total ozone (TO), which includes the ozone present in the stratospheric ozone layer and that present throughout the troposphere is measured in Dobson Units (DU) [11]. Delhi, Mumbai, Chennai, and Kolkata are four major cities of India which are located by the four regions like north, west, south, and east, respectively. The climates of these four regions are also different. This occurred due to their different geographical features. India's unique geography and geology strongly influence its climate.

In recent years, multiple linear regressions, feedforward artificial neural networks (ANN), as well as principal component regressions (combining multiple linear regressions and principal component analysis, PCA) are being used to model ozone concentrations [9, 12, 13]. The present paper deviates from earlier studies [14–16]on TO over India in the following aspects: (1) instead of trend analysis this study has intended to model TO over various urban regions of India through artificial neural network; (2) the present study has developed artificial neural network model for TO in multivariate manner, and (3) in the earlier multivariate predictive models for TO in India, the independent variables were selected based on their correlations with TO.

2   An Overview of PCA

PCA is a mathematical procedure that uses an orthogonal transformation to convert a set of observations of possibly correlated variables into a set of values of uncorrelated variables called principal components (PCs) [17]. Although PCA does not ignore covariances and correla- tions, it concentrates on variances [17]. To apply PCA, the first step is to look for a linear combination of the elements of $x = (x_1, x_2, ..., x_p)$ having maximum variance [17]

$$\alpha_1' x = \alpha_{11} x_1 + \alpha_{12} x_2 + ... + \alpha_{1p} x_p = \sum_{i=1}^{p} \alpha_{1i} x_i \tag{3}$$

Applicability of PCA to a given data set is determined by modified Bartletts sphericity test, which assumes as null hypothesis that the correlation matrix is an identity matrix. The alternative hypothesis assumes the negation of the null hypothesis. Under the assumption that the null hypothesis is true, the following chi-square (χ^2) statistic is constructed [12]

$$\chi_k^2 = \left[n - k - \frac{2(p-k)+7+2/(p-k)}{6} \right] \\ + \Sigma_{j=1}^{k} \left(\frac{\bar{\lambda}}{\lambda_j - \bar{\lambda}} \right) \times \left[-\ln \prod_{j=k+1}^{p} \lambda_j + (p-k)\ln\bar{\lambda} \right] \tag{4}$$

where p is the number of components, λ_j represents the eigenvalue for the kth component, n is the number of observations in the sample and $\bar{\lambda}$ is given by

[12]

$$\overline{\lambda} = \sum_{j=k+1}^{p} \frac{\lambda_j}{p-k} \tag{5}$$

Theoretical discussions on PCA are available in [17, 18]. Combination of PCA and ANN has been implemented in forecasting jobs in the references [12, 19, 20]. To improve interpretability, PCA is continued with a new component matrix that was created by use of Kaisers varimax rotation method in which the PC axes are rotated to a position in which the sum of variance of loading is maximum [21].

3   The Multilayer Perceptron

As multilayer perceptrons (MLPs) [22] have the potential to describe highly non-linear relationships, they are among the commonly preferred methods in atmospheric modelling studies. The MLP consists of a system of simple interconnected neurons, or nodes. The nodes are connected by weights and output signals which are a function of the sum of the inputs to the node modified by a simple nonlinear transfer, or activation function. If the transfer function was linear then the MLP would only be able to model linear functions [8, 23]. Due to its easily computed derivative a commonly used transfer function is the sigmoid function [23]

$$f(x) = \frac{1}{1+e^{-\beta x}} \tag{6}$$

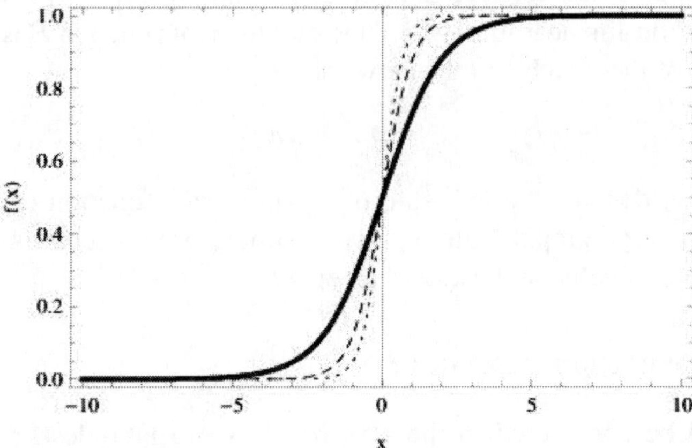

Fig. 1. The activation function $f(x) = (1+e^{-\beta x})^{-1}$. The solid, dashed and dotted lines correspond to $\beta = 1$, 2 and 3 respectively.

Form of the sigmoid function is presented in Fig. 1. MLP can learn through training. Training requires a set of training data consisting of a series of input and associated output vectors. During training the MLP is repeatedly presented with the training data and the weights in the network are adjusted until the desired input-output mapping occurs. MLPs learn in a supervised manner [8]. During training the output from the multilayer perceptron, for a given input vector, may not equal the desired output. An error signal is defined as the difference between the desired and actual output. Training uses the magnitude of this error signal to determine to what degree the weights in the network should be adjusted so that the overall error of the MLP is reduced. Let $x_p = (x_{p1}, x_{p2}, ..., x_{pm})^T \in \mathcal{R}^m$ be the pth input sample, the net input or stimulus of hidden unit h is given by [23]

$$\varphi_{ph} = \sum_{i=0}^{m} w_{hi} x_{pi} = x_p^T w_h + \theta_h \tag{7}$$

where $w_{h0} = \theta_h$ is called the hth bias. φ_{ph} is called the hth hidden basis function, which is the weighed sum of m input components. In the case of single-hidden-layer MLP $f(\varphi) = (1 + e^{-\varphi})^{-1}$ [22]. In the present work, we have constructed single-hidden-layer MLPs. The MLP is trained through BP algorithm mentioned in the Introduction. We are giving a brief overview of BP algorithm here. The standard learning scheme for the BP algorithm in which the weights of the network are updated immediately after the presentation of each pair of input and target patterns is called on-line learning [6]. An alternative training method treats all the pairs of patterns in the training set as a batch, and it updates the weights only after all training pairs in the batch have been processed, although this can result in slow convergence if the set of training patterns is large. This approach is referred to as batch learning [6]. In the present work we have adopted on-line learning, where the output error function E is a multivariate function of the weights in the network

$$E(w_k) = E_p(w_k) \tag{8}$$

where $E_p(w_k)$ denotes the half-sum-of-squares error function of the network outputs for a certain input pattern p. The learning process continues until E is less than a preset value at the end of an epoch.

4   Implementation Procedure

Variables to be considered in the present study are latitude (Lat), longitude (Long), solar zenith angle (SZA), reflectivity (REF), aerosol index (AI), sulfur dioxide index (SOI), and total ozone (TO) on daily scale. An exploration of the correlation matrices indicates the presence of multicollinearity. Kaiser-MeyerOlkin (KMO) measure has been made based on these variables to examine whether PCA is suitable for removing multicollinearity in the situation under consideration. Results are displayed in Table 1.

Study zone	KMO measure
Kolkata	0.574
Mumbai	0.552
Chennai	0.537
New Delhi	0.516

Table 1. The values of KMO for the four study zones.

It is observed in Table 1 that KMO measure is > 0.5 in each case that indicates suitability of PCA to every study zone. Now, for each study zone we compute the PCs and observe the eigen values. The PCs with eigen value > 1 are extracted. For each study zone, three such eigen values are found and in each case it is found that the three eigen values contribute roughly 75% of the total variance. In order to improve interpretability, PCA is continued with a new component matrix, i.e., created by using Kaiser varimax rotation method in which the PC axes are rotated to a position in which the sum of variance of loading is maximum. The rotated PCs are denoted as PC_R.

	K1	K2	K3	M1	M2	M3	C1	C2	C3	ND1	ND2	ND3
LAT	-0.04	0.88	-0.00	-0.01	0.90	0.08	0.93	-0.04	0.01	-0.06	0.86	-0.04
LONG	-0.04	0.89	-0.04	-0.00	0.91	0.02	0.93	-0.09	-0.02	-0.07	0.87	0.03
SZA	-0.89	0.16	-0.01	-0.88	0.25	-0.07	0.28	-0.86	0.10	-0.83	0.12	0.20
REF	0.270	0.008	-0.743	0.903	-0.008	0.009	0.020	0.865	-0.088	0.780	0.116	0.015
AI	0.43	0.08	-0.74	0.51	0.04	-0.72	0.00	0.04	0.83	0.13	0.08	0.90
SOI	0.46	-0.41	0.04	0.33	0.05	0.83	0.07	0.35	-0.74	0.42	0.12	-0.77
TO	0.89	0.07	-0.03	0.19	-0.38	0.06	0.10	0.46	0.46	0.45	-0.14	0.01

Table 2. Factor loadings for three rotated PCs (PC_R) for Kolkata (K), Mumbai (M), Chennai (C) and New Delhi (ND).

Based on the factor loadings presented in Table 2 we identify the predictors for MLP models as LAT, LONG, REF, SZA and AI. After reducing the number of the predictors by PCA, MLP models are developed to predict daily TO for all of the four study zones. sigmoid function for the hidden layer and linear functions for the output layer as activation functions. Four nodes are taken in the hidden layer to keep it free from the problem of over parameterization. To avoid the asymptotic effect, the data are standardized to $[0, 1]$ as

$$x_{\text{standardized}} = \frac{x - x_{\min}}{x_{\max} - x_{\min}} \tag{9}$$

The entire data set is divided into training and test cases in the ratio $7 : 3$. After training and testing, the model is validated over the entire set under consideration. Predictions model is assessed using Pearson correlation coefficient (PCC)

$$\rho_{xy} = \frac{Cov(x,y)}{\sigma_x \sigma_y} \tag{10}$$

and Willmott's index (WI)

$$WI = 1 - \left[\frac{\sum_{i=1}^{N}(P_i - O_i)^2}{\sum_{i=1}^{N}(|P_i - \overline{O}| + |O_i - \overline{O}|)^2} \right] \tag{11}$$

In the above equations, P_i and O_i denote the predicted and observed values respectively for the ith observation. Willmott [24] recommended computation of some 'summary measures' to assess the degree to which a model output fits an observed data set.

Study zone	WI	PCC
Kolkata	0.782	0.643
Mumbai	0.803	0.665
Chennai	0.301	0.023
New Delhi	0.254	0.245

Table 3. PCC and WI for the MLP based predictions.

The predictions are further judged by fitting trend lines to the actual-prediction pairs.

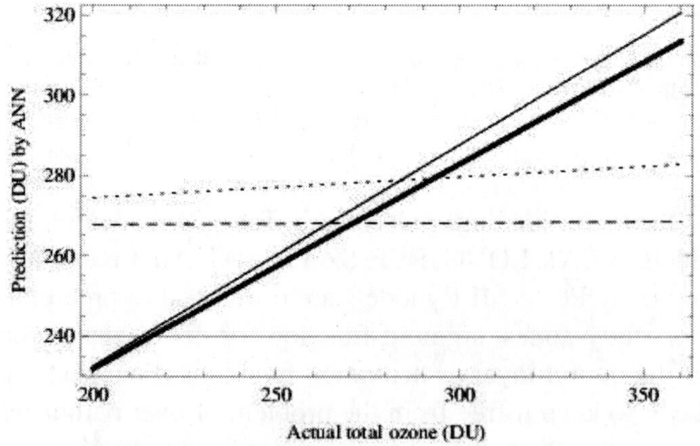

Fig. 2. The thick, opaque, dashed and dotted lines correspond to the trend lines for Kolkata, Mumbai, Chennai and New Delhi respectively.

The trend lines indicate that for Kolkata and Mumbai, the actual TO concentration could be successfully predicted by the MLP. However, for the study zones Chennai and New Delhi, the trend lines are almost horizontal and this indicates the lack of linear association between actual and ANN predicted TO concentration.

5   Concluding Remarks

In the present paper, we have considered total ozone time series over Kolkata, Mumbai, Chennai, and New Delhi, four metro cities of India using latitude, longitude, solar zenith angle, reflectivity, aerosol index, and sulfur dioxide index as the predictors. Studying the correlation matrix for all of the above study zones, it is understood that in all of the study zones, the data under consideration are characterized by multicollinearity. Implementing KMO test it is observed that for all of the above study zones, the KaiserMeyerOlkin measure is above 0.5. Thus it is understood that principal component analysis may be applied to all of the four study zones under consideration to identify the variables contributing maximum variance to the system. Single hidden layer artificial neural network in the form of multilayer perceptron is then generated for each study zone based on the predictors extracted through principal component analysis. Models of artificial neural network in the form of MLP trained through backpropagation learning have been generated for all of the study zones, and the model outcomes have been assessed statistically by means of Pearson correlation coefficient, Willmotts index. Measuring the above statistics, it has been observed that for Mumbai and Kolkata, the proposed artificial neural network model generates very good predictions. However, for Chennai and New Delhi, the predictions are not up to significant degree of accuracy. We propose to experiment with more advanced types or neural network learning for these two study zones as future study.

References

1. R. Rojas, Neural Networks: A Systematic Introduction, Springer-Verlag, Heidelberg, 1996.
2. P. S. Neelakanta, D. DeGroff, Neural Network Modeling: Statistical Mechanics and Cybernetic Perspectives, CRC Press Inc., USA, 1994.
3. S. Theodoridis and K. Koutroumbas, Pattern Recognition, Fourth Edition, Academic Press, USA, 2009.
4. T. Chen, H. Chen, R. W. Liu, Approximation capability in $C(R(n))$ by multilayer feedforward networks and related problems, IEEE Transactions on Neural Networks, 6 (1995), 25–30.
5. H. N. Mhaskar, C. H. Micchelli, Approximation by superposition of sigmoidal and radial basis functions, Advances in Applied Mathematics, 13 (1992), 350–373.
6. S. V. Kamarthi, S. Pittner, Accelerating neural network training using weight extrapolations, Neural Networks, 12 (1999), 1285–1299.

7. D. E. Rumelhart, J. L. McClelland, Parallel distributed processing: explorations in the microstructure of cognition: foundations, 1. Cambridge, MA: MIT Press, 1986.

8. M. W. Gardner, S. R. Dorling, Artificial neural networks (the multilayer perceptron)a review of applications in the atmospheric sciences, Atmospheric Environment, 32 (1998), 2627–2636.

9. S. Salcedo-Sanz, J. L. Camacho, A. M. Perez-Bellido, E. G. Ortiz-Garcia, A. Portilla-Figueras, E. Hernndez-Martn, Improving the prediction of average total ozone in column over the Iberian Peninsula using neural networks banks, Neurocomputing, 74 (2011), 1492–1496.

10. K. Andresia de Oliveira, A. Vannucci, E. Cesar da Silva, Using artificial neural networks to forecast chaotic time series, Physica A: Statistical Mechanics and its Applications, 284 (2000), 393–404.

11. S. J. Reid, Ozone and Climate Change: A Beginners Guide, Taylor and Francis, Oxford, UK, 2000.

12. S. I. V. Sousa, F. G. Martins, M. C. Pereira, M. C. M. Alvim-Ferraz, Prediction of ozone concentrations in Oporto city with statistical approaches, Chemosphere, 64 (2006), 1141–1149.

13. B. Ozbay, G. A. Keskin, S. C. Dogruparmak, S. Ayberk, Predicting tropospheric ozone concentrations in different temporal scales by using multilayer perceptron models , Ecological Informatics, 6 (2011), 242–247.

14. V. S. Tiwari, Ozone distribution over India, Pure and Applied Geophysics, 106 (1973), 1010–1017.

15. A. Sahoo, S. Sarkar, R. P. Singha, M. Kafatos, M. E. Summers, Declining trend of total ozone column over the northern parts of India, International Journal of Remote Sensing, 26 (2005), 3433–3440.

16. G. Chattopadhyay, S. Chattopadhyay, Predicting daily total ozone over Kolkata, India: skill assessment of different neural network models, Meteorological Applications, 16 (2009), 179–190.

17. I. T. Jolliffe, Principal component analysis, Springer, New York, 2002.

18. D. S. Wilks, Statistical methods in atmospheric sciences, 2nd edn. Elsevier, Oxford, 2006.

19. S. M. Al-Alawi, S. A. Abdul-Wahab, C. S. Bakheit, Combining principal component regression and artificial neural networks for more accurate predictions of ground-level ozone, Environmental Modelling and Software, 23 (2008), 396–403.

20. H.R. Maier, G.C. Dandy, Neural network based modelling of environmental variables: A systematic approach, Mathematical and Computer Modelling, 33 (2001), 669–682.

21. O. C. Karacan, Modeling and prediction of ventilation methane emissions of U.S. longwall mines using supervised artificial neural networks , International Journal of Coal Geology, 73 (2008), 371–387.

22. F. Murtagh, Neural networks for statistical and economic data: Report of the Workshop in Dublin, Ireland, December 1011, 1990, Neurocomputing, 3 (1991), 51–52.

23. G. Daqi, J. Yana, Classification methodologies of multilayer perceptrons with sigmoid activation functions, Pattern Recognition, 38 (2005), 1469–1482.

24. C. J. Willmott, Some comments on the evaluation of model performance, Bulletin of American Meteorological Society, 63 (1982), 1309–1313.

A Study on the Statefinder Description of the Interacting Ricci Dark Energy

Surajit Chattopadhyay*

Pailan College of Management and Technology, Bengal Pailan Park,
Kolkata – 700 104, West Bengal.

Abstract. Interaction between Ricci dark energy and pressureless dark matter is considered in a flat Friedman-Robertson-Walker universe. The equation of state parameter, fractional density, deceleration parameter and statefinder parameters are studied. The transition from the matter dominated decelerated to dark energy dominated accelerated phase is found achievable under this interaction.

1 Introduction

The "dark energy" (DE) that is responsible for the present accelerated expansion of the universe occupies about 70% of today's universe. Reviews on DE include [1–4]. Models of DE include quintom [1], quintessence [5], phantom [6], Chaplygin gas [7], tachyon [8], hessence [9] etc. All DE models can be classified by the behaviors of equations of state as following [1]: (i) Cosmological constant: its EoS is exactly equal to -1, that is $w_{DE} = -1$; (ii) Quintessence: its EoS remains above the cosmological constant boundary, that is $w_{DE} \geq -1$; (iii) Phantom: its EoS lies below the cosmological constant boundary, that is $w_{DE} \leq -1$ and (iv) Quintom: its EoS is able to evolve across the cosmological constant boundary. Inspired by the holographic principle [10, 11], Gao *et al.* [12] took the Ricci scalar as the IR cut-off and named it the Ricci dark energy (RDE), in which they take the Ricci scalar R as the IR cutoff. With proper choice of parameters the equation of state crosses -1, so it is a 'quintom' [13]. The Ricci scalar of FRW universe is given by $R = -6\left(\dot{H} + 2H^2 + \frac{k}{a^2}\right)$, where H is the Hubble parameter, a is the scale factor and k is the curvature. The energy density of RDE is given by $\rho_{RDE} = 3c^2\left(\dot{H} + 2H^2 + \frac{k}{a^2}\right)$. In flat FRW universe, $k = 0$ and hence $\rho_{RDE} = 3c^2\left(\dot{H} + 2H^2\right)$.

Interacting DE models have gained immense interest in recent times. Works in this direction include [14–17]. Interacting RDE was considered in [14], where the observational constraints on interacting RDE were investigated. In this work, we have considered an interacting RDE. We have reconstructed the Hubble's parameter H under this interaction and subsequently calculated the equation of state parameter w, deceleration parameter q and statefinder parameters $\{r, s\}$ in terms of redshift z.

---

* Email: surajit_2008@yahoo.co.in, surajcha@iucaa.ernet.in

2   Interacting RDE

The metric of a spatially flat homogeneous and isotropic universe in FRW model is given by

$$ds^2 = dt^2 - a^2(t) \left[dr^2 + r^2(d\theta^2 + \sin^2\theta d\phi^2) \right] \tag{1}$$

where $a(t)$ is the scale factor. The Einstein field equations are given by

$$H^2 = \frac{1}{3}\rho \tag{2}$$

and

$$\dot{H} = -\frac{1}{2}(\rho + p) \tag{3}$$

where ρ and p are energy density and isotropic pressure respectively (choosing $8\pi G = c = 1$).

The conservation equation is given by

$$\dot{\rho} + 3H(\rho + p) = 0 \tag{4}$$

As we are considering interaction between RDE and dark matter, the conservation equation will take the following form

$$\dot{\rho}_{total} + 3H(\rho_{total} + p_{total}) = 0 \tag{5}$$

where, $\rho_{total} = \rho_{RDE} + \rho_m$ and $p_{total} = p_{RDE}$ (as we are considering pressureless dark matter, $p_m = 0$). As in the case of interaction the components do not satisfy the conservation equation separately, we need to reconstruct the conservation equation by introducing an interaction term Q. Considering the interaction term Q as $Q = 3H\delta\rho_m$ [18, 19], where δ is the interaction parameter, the conservation equation (5) takes the form

$$\dot{\rho}_{RDE} + 3H(\rho_{RDE} + p_{RDE}) = Q \tag{6}$$

and

$$\dot{\rho}_m + 3H\rho_m = -Q \tag{7}$$

From equations (3) and (7) we express p_{RDE} under interaction in a flat FRW universe as

$$p_{RDE} = -\left[(2 + 3c^2)\dot{H} + 6c^2H^2 + \rho_{mo}a^{-3(1+\delta)} \right] \tag{8}$$

Using the energy density and pressure of RDE in (6) we express $H(z)$ as

$$H(z)^2 = \frac{c^2}{-1 + 2c^2} + B_0(1+z)^{-\frac{2}{c}+4c} + \frac{2c(1+z)^{3(1+\delta)}\rho_{mo}}{6 + 3c(3 - 4c + 3\delta)} \tag{9}$$

Using the above form of Hubble's parameter in p_{RDE} and ρ_{RDE} we get the pressure and energy density of RDE under interaction. Subsequently we calculate the equation of state parameter $w_{RDE} = \frac{p_{RDE}}{\rho_{RDE}}$ and plot in Fig. 1 against redshift z for various values of c^2.

Also, we derive the fractional densities for dark energy and dark matter

$$\Omega_{de} = \frac{\rho_{DE}}{3\tilde{H}^2} \; ; \qquad \Omega_m = \frac{\rho_m}{3\tilde{H}^2} \tag{10}$$

where, $\tilde{H} = H(z)$. Fractional densities obtained this way are plotted in Fig. 2.

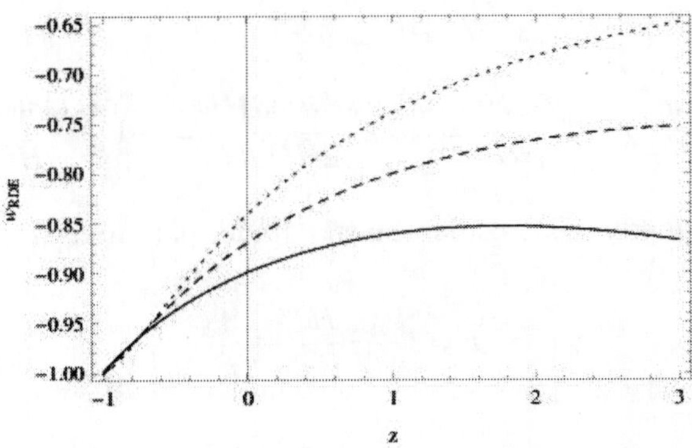

Fig. 1. Variations of equation of state parameter w_{RDE} against redshift z for $\delta = 0.05$ and $\rho_{mo} = 0.23$. The solid, dashed and dotted lines correspond to $c^2 = 0.7$, 0.75 and 0.8 respectively.

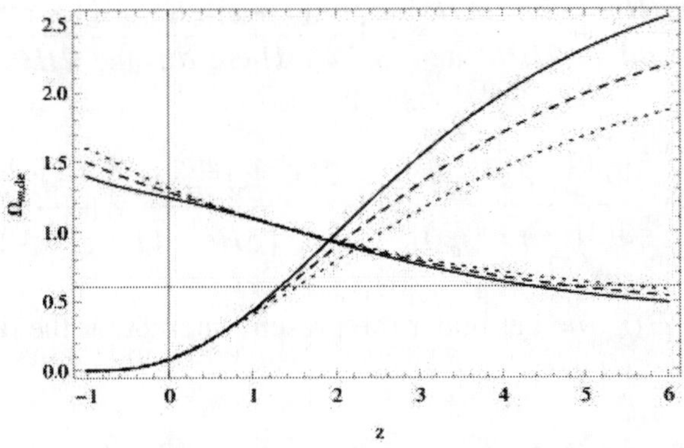

Fig. 2. Evolution of fractional densities $\Omega_{de} = \frac{\rho_{RDE}}{3\tilde{H}^2}$ and $\Omega_m = \frac{\rho_m}{3\tilde{H}^2}$ with z for different values of c^2. The Ω_m is decaying from the very early stage to the later stages and Ω_{de} is gradually increasing.

3   Statefinder Diagnostics

The problem of discriminating different dark energy models is now emergent. In order to solve this problem, a sensitive and robust diagnostic for dark energy is a must. The statefinder parameter pair $\{r,s\}$ introduced by [20] is proven to be useful tools for this purpose. The statefinder pair is a 'geometrical' diagnostic in the sense that it is constructed from a space-time metric directly, and it is more universal than 'physical' variables which depends upon properties of physical fields describing dark energy, because physical variables are, of course, model-dependent [13]. First, we consider deceleration parameter q [21, 22].

$$q = -\frac{a\ddot{a}}{\dot{a}^2} = -1 - \frac{\dot{H}}{H^2} = -1 - \frac{a}{2\tilde{H}^2}\frac{d\tilde{H}^2}{da} = -1 + \frac{(1+z)}{\tilde{H}^2}\frac{d\tilde{H}^2}{dz} \qquad (11)$$

and subsequently we consider the statefinder parameters

$$r = 1 + 3\frac{\dot{H}}{H^2} + \frac{\ddot{H}}{H^3} \qquad (12)$$

and

$$s = -\frac{3H\dot{H} + \ddot{H}}{3H(2\dot{H} + 3H^2)} \qquad (13)$$

Or, equivalently [21]

$$r = 1 + \frac{2a}{\tilde{H}^2}\frac{d\tilde{H}^2}{da} + \frac{a^2}{2\tilde{H}^2}\frac{d^2\tilde{H}^2}{da^2} = 1 - \frac{(1+z)}{\tilde{H}^2}\frac{d\tilde{H}^2}{dz} + \frac{(1+z)^2}{2\tilde{H}^2}\frac{d^2\tilde{H}^2}{dz^2} \qquad (14)$$

and

$$s = -\frac{4a\frac{d\tilde{H}^2}{da} + a^2\frac{d^2\tilde{H}^2}{da^2}}{3(3\tilde{H}^2 + a\frac{d\tilde{H}^2}{da})} = -\frac{2(1+z)\frac{d\tilde{H}^2}{dz} - (1+z)^2\frac{d^2\tilde{H}^2}{dz^2}}{3\left(3\tilde{H}^2 - (1+z)\frac{d\tilde{H}^2}{dz}\right)} \qquad (15)$$

Using $\tilde{H} = H(z)$ we get under the present interaction the deceleration parameter as

$$q(z) = -1 + \frac{\frac{2(-1+2c^2)B_0(1+z)^{-\frac{2}{c}+4c}}{c} + \frac{6c(1+z)^{3(1+\delta)}(1+\delta)\rho_{mo}}{6+3c(-4c+3(1+\delta))}}{\frac{c^2}{2c^2-1} + B_0(1+z)^{-\frac{2}{c}+4c} + \frac{2c(1+z)^{3(1+\delta)}\rho_{mo}}{6+3c(-4c+3(1+\delta))}} \qquad (16)$$

The deceleration parameter is plotted against z in Fig. 3 and the $\{s-r\}$ trajectory is plotted in Fig. 4 for different values of c^2 in Fig. 4.

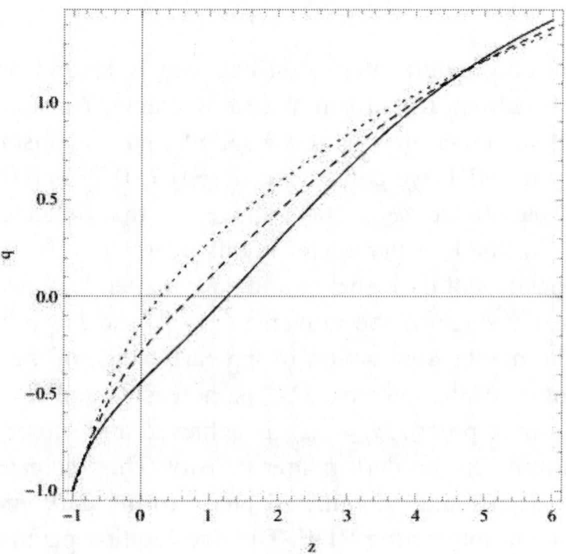

Fig. 3. Plotting of the deceleration parameter q against z for $\delta = 0.05$ and $\rho_{mo} = 0.23$. The solid, dashed and dotted lines correspond to $c^2 = 0.7,\ 0.75$ and 0.8 respectively.

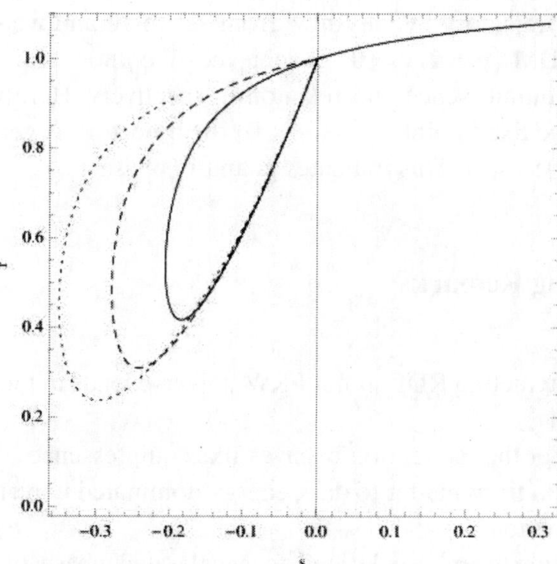

Fig. 4. Plotting of the $\{s - r\}$ trajectory for $\delta = 0.05$ and $\rho_{mo} = 0.23$. The solid, dashed and dotted lines correspond to $c^2 = 0.7,\ 0.75$ and 0.8 respectively.

4   Discussions

In this work we considered interaction between RDE and pressureless dark matter in a flat Friedman-Robertson-Walker universe. We studied the EoS parameter under this interaction ($\delta = 0.05$) and plotted it against redshift z in Fig. 1. In the density for RDE we considered $c^2 = 0.7$, 0.75 and 0.8. In Fig. 1 we found that EoS parameter ω_{RDE} stays above -1 that indicates quintessence-like behavior [1] of the EoS parameter in this interaction. Also, we considered the fractional densities of dark energy and dark matter. It is observed from Fig. 2 that at very early stage of the universe ($z > 2$) the Ω_m is dominating Ω_{de} that indicates the matter domination in the early stage of the universe. However, with evolution of the universe Ω_{dm} is increasing and Ω_m in decaying. At $z \approx 1.7$ the crossover point $\Omega_{de} = \Omega_m$ is achieved and subsequently dark energy density dominated the dark matter density. This indicates the evolution of the universe from matter dominated phase to the dark energy dominated phase in presence of interacting RDE. The deceleration parameter q is viewed against redshift z in Fig. 3, where it is observed that irrespective of the values of c^2 under consideration the universe is transiting from the decelerated to the accelerated phase ($q > 0$ to $q < 0$)in presence of this interaction. However, we observe that as we are increasing the value of c^2 the point of transition is changing from higher to lower values of z. This indicates that increase in c^2 is delaying the transition from deceleration to acceleration. In Fig. 4 we have created $\{r - s\}$ trajectory by varying z from -1 to 10 and we observe that the fixed point ΛCDM ($r = 1, s = 0$) is achieved. We know that $s > 0$ and $s < 0$ corresponds to quintessence and phantom respectively. Here we observe that after crossing the fixed point ($r = 1, s = 0$) the trajectory is getting confined in the region, where $s < 0$. This indicates phantom phase.

5   Concluding Remarks

The study on interacting RDE in flat FRW universe leads to the following crisp conclusions:
(i) The EoS under the interaction behaves likes quintessence.
(ii) The transition from matter to dark energy dominated universe is achievable under this interaction.
(iii) The transition from decelerated to accelerated phase of the universe is possible under this interaction. However, increase in the value of c^2 delays the transition from deceleration to acceleration.
(iv) The ΛCDM phase is obtainable under this interaction. However, the $\{r - s\}$ trajectory can go beyond this fixed point.

Acknowledgment

The author is thankful to the organizers of CRDMC-2012 for kindly accepting this work for presentation.

References

1. Y-F. Cai, E.N. Saridakis, M.R. Setare, J-Q. Xi, Quintom cosmology: theoretical implications and observations, Physics Reports, 493 (2010), 1–60.
2. E.J. Copeland, M. Sami, S. Tsujikawa, Dynamics of dark energy, International Journal of Modern Physics D, 15 (2006), 1753–1936.
3. T. Padmanabhan, Dark energy: the cosmological challenge of the millennium, Current Science, 88 (2005), 1057–1067.
4. M. Li, X-D. Li, S. Wang, Y. Wang, Dark energy, Communications in Theoretical Physics, 56 (2011), 525–604.
5. B. Ratra, P.J.E. Peebles, Cosmological consequences of a rolling homogeneous scalar field, Physical Review D, 37 (1988), 3406–3427.
6. S. Nojiri, S.D. Odintsov, Unifying phantom inflation with late-time acceleration: scalar phantom non-phantom transition model and generalized holographic dark energy, General Relativity and Gravitation, 38 (2006), 1285–1304.
7. V. Gorini, A. Kamenshchik, U. Moschella, Can the Chaplygin gas be a plausible model for dark energy? Physical Review D, 67 (2003), 063509.
8. L.P. Chimento, Extended tachyon field, Chaplygin gas, and solvable k-essence cosmologies, Physical Review D, 69 (2004), 123517.
9. W. Zhao, Holographic hessence models, Physics Letters B, 655 (2007), 97–103.
10. S.D. Hsu, Entropy bounds and dark energy, Physics Letters B, 594 (2004), 13–16.
11. X. Xhang, Holographic Ricci dark energy: current observational constraints, quintom feature, and the reconstruction of scalar-field dark energy, Physical Review D, 79 (2009), 103509.
12. C. Gao, F. Wu, X. Chen, Y-G. Shen, Holographic dark energy model from Ricci scalar curvature, Physical Review D, 79 (2009), 043511.
13. C-J. Feng, Statefinder diagnosis for Ricci dark energy, Physics Letters B, 670 (2008), 231–234.
14. M. Suwa, T. Nihei, Observational constraints on the interacting Ricci dark energy model, Physical Review D, 81 (2010), 023519.
15. M.R. Setare, E. C. Vagenas, Thermodynamical interpretation of the interacting holographic dark energy model in a non-flat universe, Physics Letters B, 666 (2008), 111–115.
16. M.R. Setare, Interacting generalized Chaplygin gas model in non-flat universe, European Physical Journal C, 52 (2007), 689–692.
17. M.R. Setare, J. Sadeghi, A.R. Amani, Interacting tachyon dark energy in non-flat universe, Physics Letters B, 673 (2009), 241–246.
18. A. Sheykhi, Interacting agegraphic tachyon model of dark energy, Physics Letters B, 682 (2010), 329–333.
19. M. Cataldo, P. Mella, P. Minning, J. Saavedra, Interacting cosmic fluids in power-law FriedmannRobertsonWalker cosmological models, Physics Letters B, 662 (2008), 314–322.
20. V. Sahni, T.D. Saini, A.A. Starobinsky, Statefinder - A new geometrical diagnostic of dark energy, JETP Letters, 77 (2003), 201–206.
21. U. Debnath, S. Chattopadhyay, Statefinder and Om diagnostics for interacting new holographic dark energy model and generalized second law of thermodynamics, arXiv:1102.0091v2 [physics.gen-ph], 2011.
22. S. Chattopadhyay, U. Debnath, G. Chattopadhyay, Acceleration of the Universe in presence of tachyonic field, Astrophysics and Space Science, 314 (2008), 41–44.

Dynamical Formalism of Diffeomorphism Invariance and Loop Quantum Gravity

L. Mullick

Department of Mathematics, Hiralal Mazumdar Memorial College For Women
Dakshineswar, Kolkata – 700 035, West Bengal.

Abstract. It is argued that black hole evaporation can be realized as an effect of quantum geometry using a dynamical formalism of diffeomorphism invariance. This formalism suggests that torsion acts within a quantized area unit associated with a loop and this forbids the Hamiltonian constraint to be satisfied for a finite loop size. This helps us to assign a spin in each area bit of the event horizon of a black hole so that it forms a spin system.

1 Introduction

The necessity of a quantum theory of gravity was pointed out by Einstein in 1916 in a paper in the *Preussische Akademie Sitzungsberichte*.

He wrote: *Nevertheless, due to the inneratomic movement of electrons, atoms would have to radiate not only electromagnetic but also gravitational energy, if only in tiny amounts. As this is hardly true in nature, it appears that quantum theory would have to modify not only Maxwellian electrodynamics but also the new theory of gravitation.*

Papers on the subject began to appear in the thirties most notably by Bronstein, Rosenfeld and Pauli. However, detailed work began only in sixties.

Quantum theory seems to be a universal theory of Nature. It provides a general framework for all theories describing particle interactions. The only interaction that has not been fully accomodated within quantum theory is the gravitational field, the oldest known interaction. It is very successfully described by a classical theory, Einstein's general theory of relativity (GR)

The classical Einstein equation is given by

$$R_{\mu\nu} - \frac{1}{2}Rg_{\mu\nu} = \kappa T_{\mu\nu}(g) \tag{1}$$

These equations relate matter density in form of the energy-momentum tensor $T_{\mu\nu}$ and geometry in form of the Ricci curvature tensor $R_{\mu\nu}$. The metric tensor $g_{\mu\nu}$ enters into the definition of energy-momentum tensor. However, while the left hand side is described by a classical theory, the right hand side is governed by a quantum field theory (QFT). Hence we are enforced to quantize the metric itself, that is, we need a quantum theory of gravity.

Here three types of quantization, namely, covariant quantization, canonical quantization, loop quantization will be discussed.

2 Covariant Quantization

Renowned Physicist Feynman in 1963 took this idea first of treating gravity on an equal footing with other field theories. The covariant approach tries to solve this quantization problem from the point of view of particle physicist by mimicking the method for electromagnetic fields. The covariant quantization of gravitation field deals with a Lorentzian manifold $(M, g^B_{\mu\nu})$ where $g^B_{\mu\nu}$ is a Lorentzian metric generally called the background metric over a differentiable manifold M [1]. The components of the quantum metric field $g^Q_{\mu\nu}$ are then defined by

$$g^Q_{\mu\nu} = g^B_{\mu\nu}(x) + h_{\mu\nu}(x), \quad x \in M \tag{2}$$

where μ, $\nu = 0, 1, 2, 3$ and $h_{\mu\nu}(x)$ is viewed as quantum fluctuation and it contains the non-linearities of the metric tensor. Now one could use the familiar perturbation technique of quantum field theories to obtain a formal perturbation series [2]. In the context of quantum gravity, the term 'covariant' is somwhat misleading because the introduction of a background metric violates diffeomorphism covaiance. It is used mainly to emphasize that this approach does not involve a $3+1$ decomposition of spacetime. When equation (2) is substituted in Einstein Lagrangian

$$\tilde{\mathscr{L}}_E = (2\kappa)^{-1}\sqrt{-g}R \tag{3}$$

a nonlinear interaction is obtained and the perturbation theory generated by $h_{\mu\nu}(x)$ is not renormalizable [2].

3 Canonical Quantization

The canonical quantization scheme can be used for a theory which can be cast in the Hamiltonian form. The canonical quantization scheme provides a natural avenue in this direction since it does not require the fixation of classical background geometry. The outlines of the canonical quantization seem straight forward. First, the canonical theory is cast into Hamiltonian form and the pairs of canonically conjugate variables are identified.

The special feature of the gravitational field is that it does not naturally lead to a Hamiltonian but only to a Hamiltonian constraint. The circumstance incredibly complicates the implementation of the canonical programme. In the constrained Hamiltonian formulation of general relativity, as developed by Dirac and others the basic variables are the three-metric on a spatial slice and the extrinsic curvature. Both of these have a direct geometrical interpretation, but these choice of variable leads to complicated, nonpolynomial constraints, which no one has been able to consistently carry over to the quantum theory.

Recently starting from the Hamiltonian formulation of gravity in the triad formalism Ashtekar [3] in 1986 has performed a canonical transformation to a new set of variables, which leads to a dramatic simplification of the theory of constraint.

In a three plus one splitting of space-time we take as basic configuration variables

$q_{ij}(i, j = 1, 2, 3)$ the metric of the three-space sections. One needs to know not only the intrinsic geometry of the three space but also the way it is embedded in the four dimensional space-time. For this we need the spatial part of the covariant derivative of the normal to the three-space, since this tells us how a given three-space is bent in four-dimensions. Thus the variables $K_{ij} = n_{i;j}$ where ; means the covariant derivative with respect to the four-space metric $^4g_{\mu\nu}$ ($\mu, \nu = 0, 1, 2, 3$). The canonical approach always starts with the generalization of time. Let us cut the space-time by an arbitrary spacelike hypersurface Σ, $X^\alpha = X^\alpha(x^a)$ where $a = 1, 2, 3$ and $\alpha = 0, 1, 2, 3$. At each point of the hypersurface, a basis consisting of three tangent vectors X_a^α to the hypersurface and of unit normal vector n^α. Let us now foliate space-time by deforming the hypersurface Σ in a continuous way, that is to decompose spacetime in $3+1$ which implies $\mathcal{M} = R \times \Sigma$. This gives us a one-parameter family of hypersurfaces $X^\alpha = X^\alpha(x^a, t)$. The deformation vector

$$N^\alpha = \dot{X}^\alpha = \frac{\partial X^\alpha(x^a, t)}{\partial t} \tag{4}$$

connecting the points with the same label x^a on two neighbouring hypersurfaces can be decomposed with respect to the basis vectors $\{n^\alpha, X_a^\alpha\}$,

$$N^\alpha = Nn^\alpha + N^a X_a^\alpha \tag{5}$$

Its components N and N^a are called the lapse and the shift functions by Arnowitt, Deser and Misner (1962) [4]. Their physical interpretation stems from the $(1+3)$ form of writing the space-time metric.

$$ds^2 = -(N\,dt)^2 + q_{ab}(dx^a + N^a dt)(dx^b + N^b dt)\ (a,\ b = 1, 2, 3) = g_{\mu\nu}dx^\mu dx^\nu \tag{6}$$

From this the components of $g_{\mu\nu}$ can be derived.

$$g_{00} = q_{ab}N^a N^b - N^2$$

$$= N_a N^a - N^2$$

$$g_{0j} = q_{ij}N^i = N_j$$

$$g_{ij} = q_{ij}$$

The vector $(N, N^1, N^2, N^3)\,\delta t$ connecting the point (t, x^i) with the point $(t+\delta t, x^i)$, $N\delta t$ specifies a proper displacement normal to the surface $x^0 = t$, N is called the lapse function, $N^i\delta t$ gives the displacement from the point (t, x^i)

144   L. Mullick

to the foot of the normal to $x^0 = t$ through $(t + \delta t, \, x^i)$, thus N^i is named as shift function.

Arnowitt, Deser and Misner [4] found that a better choice of variables than $K^{ab}(a, \, b = 1, \, 2, \, 3)$.

$$p^{ab} = (det \, q)^{1/2} G^{-1}(K^{ab} - K^{mn} q_{mn} q^{ab}) \tag{7}$$

where G is Einstein's constant $[G = 8\pi c^{-3}$, Newton's constant$]$ q^{ab} is the matrix inverse of q_{ab} and

$$\frac{\partial q_{ab}}{\partial t} = \frac{2N}{\sqrt{q}}\left(p_{ab} - \frac{1}{2} q_{ab} p_k^k\right) + N_{a|b} + N_{b|a} \tag{8}$$

where $|$ denotes the covariant derivative in the 3-space and $p_i^i = p_{ab} q^{ab}$.

In this respect an exceedingly convenient new set of variables were introduced by A. Ashtekar [3]. For simplicity he has considered compact hypersurface Σ only.

The phase space $\tilde{\Gamma}$ which is a pair (q_{ab}, p^{ab}) satisfying

$$q_{ab} = \left[1 + \frac{M(\theta, \phi)}{r}\right]^4 e_{ab} + O\left(\frac{1}{r^2}\right)$$

$$p^{ab} q_{ab} = O\left(\frac{1}{r^3}\right)$$

$$p^{ab} - \frac{1}{3} p q^{ab} = O\left(\frac{1}{r^2}\right) \tag{9}$$

where e_{ab} is a fixed positive-definite metric on Σ which is Euclidean i.e., flat outside some compact set, r denote a radial coordinate with respect to e_{ab}.

This space is extended in order to incorporate spinor fields. In addition to tensor fields $T^{a...b}_{c...d}$ on Σ, objects such as $\lambda^{A...B}_{M...N}\,^{a...b}_{c...d}$ with internal $SU(2)$ indices $A...B, \; M...N$ are considered. These internal indices are not to be thought of as spinor indices, since we do not yet have soldering form σ^B_{aA} to tie them down to the tangent space of Σ. Isomorphisms σ^B_{aA} from the tangent vectors λ^a to Σ to the trace-free Hermitian, second rank spinors $\lambda^B_A : \lambda^B_A = \sigma^B_{aA} \lambda^a$ are introduced. Now given a specific σ, we are back to the standard spinorial scenario. The metric q_{ab} is now thought of as a secondary object, derived from the primary dynamical variable σ^B_{aA} via

$$q_{ab} = -\sigma^B_{aA} \sigma^A_{bB} \equiv -Tr\sigma_a \sigma_b \tag{10}$$

There exists a preferred nowhere vanishing skew field ε^{AB}, denoting its inverse by ε_{AB}

$$\varepsilon^{AB} \varepsilon_{AM} = \delta^B_M \; and$$

$$\lambda^A = \varepsilon^{AB} \lambda_B \; and \; \mu_A = \varepsilon_{BA} \mu^B \tag{11}$$

Now the extended phase space Γ will be obtained by fixing a soldering form $^0\sigma_{aA}^{\ B}$ outside some compact region in Σ whose connection D is flat. Thus $^0\sigma$ is a soldering form of an Euclidean metric e_{ab}. $\mathscr{C}$ is denoted by the space of all soldering forms $\sigma_A^{a\ B}$ (inverse mapping of σ_{aA}^B) such that

$$\sigma_A^{a\ B} = \left[1 + \frac{M(\theta,\phi)}{r}\right]^2 {}^0\sigma_A^{a\ B} + O\left(\frac{1}{r^2}\right) \tag{12}$$

$\mathscr{C}$ is the new configuration space. Given any σ in $\mathscr{C}$ we obtain a q in $\tilde{\mathscr{C}}$ via (10). Thus there is a natural projection ψ from the new configuration space $\mathscr{C}$ to $\tilde{\mathscr{C}}$: $\psi(\sigma_A^a{}^B) = q_{ab}$, where q_{ab}. Then it follows from (10) that σ_1 and σ_2 are related by a local $SU(2)$ transformation. Thus the enlargement of the configuration space from $\tilde{\mathscr{C}}$ to $\mathscr{C}$ has been brought about because of the freedom to perform internal $SU(2)$ rotations. While q_{ab} has six components per space point, $\sigma_A^{a\ B}$ has nine, the new three degrees of freedom correspond to precisely the three $SU(2)$ rotations. This means that we now have three new constraints.

The momentum conjugate to $\sigma_A^{a\ B}$ is taken as M_{aA}^B, now fix a point (σ, M) of Γ. Then two connections are introduced, $^{\pm}\mathscr{D}$ which act on tensor and spinor fields on (Σ, σ).

$$^{\pm}\mathscr{D}\lambda_{bM} = D_a\lambda_{bM} \pm \frac{i}{\sqrt{2}}\pi_{aM}^N\lambda_{bN} \tag{13}$$

where D_a is the connection and π_{aM}^N is given by

$$\pi_{aM}^N = G(det\ q)^{-1/2}\left[M_{aM}^N + \frac{1}{2}(Tr\ M_b\sigma^b)\sigma_{aM}^N\right] \tag{14}$$

In order to stress the analogy with ordinary gauge theories, we write

$$^{\pm}\mathscr{D}\lambda_M = \partial_a\lambda_M \pm G^{\pm}A_{aM}^N\lambda_N \tag{15}$$

so that (13) yields

$$G^{\pm}A_{aM}^N = \Gamma_{aM}^N \pm \frac{i}{\sqrt{2}}\pi_{aM}^N \tag{16}$$

where Γ_{aM}^N are the spin connection one-forms $(D_a - \partial_a)\lambda_M \equiv \Gamma_{aM}^N\lambda_N$ and

$$\tilde{\sigma}_A^{a\ B} = (det\ q)^{1/2}\sigma_A^{a\ B} \tag{17}$$

Thus $\tilde{\sigma}$ may be thought of as being canonically conjugate to $^{\pm}A_a$. The basic new variables of Ashtekar are $(\tilde{\sigma}_A^{a\ B}, {}^+A_{aA}^B)$ or $(\tilde{\sigma}_A^{a\ B}, {}^-A_{aA}^B)$.

The constraints are

$$Tr\ \tilde{\sigma}^a\tilde{\sigma}^b\ {}^{\pm}F_{ab} = 0 \tag{18}$$

$$Tr\sigma^{a\pm}F_{ab} = 0 \tag{19}$$

$$M_{[ij]} = 0 \tag{20}$$

which is equivalent to

$$^{\pm}\mathscr{D}_a \tilde{\sigma}_A^{a\,B} = 0 \tag{21}$$

where

$$^{\pm}F_{abM}^{N} = \partial_{[a}{}^{\pm}A_{b]M}^{N} + G\left[{}^{\pm}A_a, {}^{\pm}A_b\right]_M^N$$

We observe that the remarkable fact that the constraints are now, at the worst quadratic in each of the basic variables. The whole system of constraints bears an amazing analogy with the corresponding one in a non-Abelian gauge theory and the constraints can be cast into Yang-Mills form. The connection one-forms $^{\pm}A_{aA}^{B}$ can be thought of Yang-Mills connection one-forms on the three manifold Σ, and its curvature, $^{\pm}F_{abA}^{B}$, as the dual of the magnetic field $^{\pm}B_A^{m\,B}$, i.e.,

$$^{\pm}B_A^{m\,B} = \varepsilon^{mab}{}^{\pm}F_{abA}^{B}$$

Then the constraints (18), (19) and (20) become

$$Tr\,\varepsilon(E \cdot E \times B) = 0 \tag{18$\prime$}$$

$$Tr\,(E \times B) = 0 \tag{19$\prime$}$$

$$^{\pm}\mathscr{D}_a\,E_A^{a\,B} = 0 \tag{20$\prime$}$$

So in this way Ashtekar greatly simplified the constraints to polynomial constraints. The Hamiltonian is a linear combination of the constraints. Therefore general relativity can be viewed as Yang-Mills theory with some extra constraints.

Unfortunately there was one choice that was made in obtaining the polynomial constraint equations that shows to be the end of the connection representation. The Barbero-Immirzi [5] parameter γ, which was chosen equal to i, should be a real number. If it is kept a complex number the phase space of general relativity is now in the complex plane. This imposes the challenge of finding reality conditions after quantizing. Finding suitable reality conditions for a complex theory of quantum gravity proved to be impossible. Therefore the complex value of γ was dropped and with that also the polynomial form of the hamiltonian constraint.

4 Loops

With the introduction of the formulation of the triads we came across a new constraint. This is the $SO(3)$-invariance. There is a large class of functionals in terms of the Ashtekar's variables that is already $SO(3)$ invariant. These are the Wilson loops, the trace of the holonomy of the Ashtekar variable.

$$W_\gamma(A) = Tr\left(P\,exp\oint dy^a A_a\right) \tag{22}$$

where $\gamma : [0,1] \longrightarrow \Sigma$.

These loops form a basis for all $SO(3)$-invariant functionals. These loops can be taken as our basic variables. This is called loop representation. Now what happens to the hamiltonian and diffeomorphism constraint equations with the Wilson loops as our new canonical variables? First the diffeomorphism constraint seemed to be solved naturally. It also seemed that for smooth loops the hamiltonian constraint was satisfied. Loop quantum gravity is seemingly less plagued by a lack of predictions, and indeed it is often claimed that the discreteness of area and volume operators are concrete predictions of the theory.

Here a unified picture of quantum gravity with the quantization procedure of fermions in the framework of the relativistic generalization of stochastic quantization procedure is developed.

In an earlier paper [6] it has been pointed out that the quantization of a fermion can be achieved when we introduce a direction vector ξ_μ attached to a space-time point x_μ which appears as an internal variable giving rise to the internal degree of freedom representing spin. It has been shown that the complexified coordinate system $z_\mu = x_\mu + i\xi_\mu$ where ξ_μ corresponds to a direction vector effectively gives rise to the $SL(2,C)$ gauge theoretical extension of the space-time coordinates. In this manifold the position and momentum variables can be written as

$$Q_\mu = -i\left(\frac{\partial}{\partial p_\mu} + \mathscr{A}_\mu(p)\right)$$

$$P_\mu = i\left(\frac{\partial}{\partial q_\mu} + \mathscr{A}_\mu(q)\right) \tag{23}$$

where $\mathscr{A}_\mu \in SL(2,C)$ [6–8] and q_μ (p_μ) represents the mean position of the space-time (momentum) coordinates. In three space dimensional manifold Σ we may take the compact group $SU(2)$. In this formalism spin is represented as an $SU(2)$ gauge bundle [9].

It is observed that the gauge theoretically extended space-time as given by $z_\mu = x_\mu + i\xi_\mu$ where ξ_μ related to the quantization of a fermion essentially implies that a massive fermion is an extended body and corresponds to a skyrmion. The overall space where a direction vector (vortex line) which is topologically equivalent to a magnetic flux line attached to a space point effectively corresponds to the discretization of space depicting a lattice structure. This follows from the fact that any two space points cannot come close together within an infinitesimal distance due to the vortex lines. In fact as two vortices cannot coincide at a point we will have patches of space and the space made of homogeneous patches defines a lattice. As this manifold is related to the quantization of a fermion where the direction vector plays the role of an internal degree of freedom corresponding to spin, we may take the size of the direction vector to be of the order of Planck length. It is noted that when a direction vector (vortex line) is attached to a space point, it specifies a particular orientation corresponding to a fixed angular deviation from the quantization axis (z-axis). This means that we have specific background here and represents a specific lattice structure of the space. However if the direction vector (vortex line) is rotated around the z-axis having a full 2π rotation the system becomes background independent and corresponds to a lattice theory where we consider all lattice structures. This means that we have now diffeomorphism invariance. Indeed when we have a full 2π rotation of the direction vector the angle of deviation of this vector from the z-axis is not fixed but covers all the possible angles and hence all possible backgrounds. Thus the system becomes background independent and leads to the realization of diffeomorphism invariance. It effectively creates a closed area on the surface of the three dimensional space giving rise to a loop. Thus we note that when background independence and diffeomorphism invariance appears as an emergent symmetry in this way, the formation of a loop and its specific area determines the characteristic feature of such a theory.

Given a loop γ and a connection $\mathscr{A}$ we can form the holonomy $Pe^{\int_\gamma \mathscr{A}}$ which represents the path integral exponential around a closed loop γ and P denotes the path ordering along γ. The gauge invariant quantity in a non-Abelian gauge theory is the Wilson loop which is the trace of the holonomy

$$T(\gamma, \mathscr{A}) = TrP\, e^{\int_\gamma \mathscr{A}} \tag{24}$$

It is noted that in three space dimensional manifold Σ where the direction vector traverses an area in the surface by rotating about the z-axis we can visualize this as if the non-Abelian magnetic field associated with the vector potential given by (23) is rotating around the z-axis with a certain angular velocity. This will induce a non-Abelian electric field around the area traversed by it.

The conjugate momenta of the connection $\mathscr{A}_a^i$ is given by the electric field E_a^i where $a = 1, 2, 3$ are space indices and $i = 1, 2, 3$ are group indices. The electric field as a density is equivalent to a two-form

$$E_{ab}^i = \varepsilon_{abc}\widetilde{E}^{ci} \tag{25}$$

The integrated form of this against any surface $S \in \Sigma$ gives rise to the electric flux

$$E^i(S) = \int_S d^2 S^{ab} E_{ab}^i \tag{26}$$

It is noted that the electric flux measured by $E^i(S)$ is quantized. The vacuum state is an eigenstate of electric flux with vanishing flux everywhere

$$E(S)\,|\,0\rangle = 0 \tag{27}$$

for all surfaces. We can create a state which has one unit of non-Abelian electric flux around a loop by acting on the vacuum with a Wilson loop operator

$$T(\gamma,\mathscr{A})\,|\,0\rangle = |\,\gamma\rangle \tag{28}$$

We can consider a gauge invariant version of the electric flux

$$\mathscr{E}(S) = \int_S \sqrt{E^i E^i} \tag{29}$$

where the flux is defined here in a diffeomorphism invariant way. Thus we can define a unit area of the surface

$$Area = \hbar \mathscr{G} \mathscr{E}(S) \tag{30}$$

The states created by the Wilson loops acting on the vacuum $\,|\,0\,\rangle\,$ are eigenstates of the operators that measure the areas of surfaces [10–21]. The factor $\hbar\mathscr{G}$ implies that the minimal quantum of area is given by the Planck area $\hbar\mathscr{G}$. Thus we observe that when diffeomorphism invariance is introduced in the quantum geometry associated with the quantization procedure of a fermion we have the generation of loops and quantized area bits as a natural consequence.

We have pointed out that in the framework of quantization of a fermion the introduction of diffeomorphism invariance through the rotation of the direction vector attached to a spatial point leads to the generation of the quantized area loop. Also as the direction vector gives rise to the spin degrees of freedom, each area bit is characterized by the fact that there is a spin inserted in it. Let us take the space-time manifold $M = R \times \Sigma$ where Σ is the 3-space manifold. We now consider that the area bits in the surface of the 3-space manifold are arranged in such a way that the total spin-spin interaction minimizes.

In quantum gravity area and volume will appear as operators operating on the spin network states. The important result in LQG is that with $j_i \in Z/2$

$$A(\Sigma)\psi_s(\mathscr{A}) = \hbar\mathscr{G} \sum_i \sqrt{j_i(j_i+1)}\;\psi_S(\mathscr{A}) \tag{31}$$

where i labels the puncture and $A(\Sigma)$ defines the area operator acting on the functional $\psi_s(\mathscr{A})$ depicting the spin network state. This follows from the definition of area unit given by Eqn (31) where the electric flux corresponds to the conjugate of the gauge potential $-i\frac{\delta}{\delta\mathscr{A}_a^i(x)}$. Indeed we have the relation

$$E(\Sigma)\psi_S(\mathscr{A}) = \sqrt{j(j+1)}\;\psi_S(\mathscr{A}) \tag{32}$$

so that we can write for the eigenvalue of the area operator as

$$A(\Sigma) = \hbar\mathscr{G}\,\gamma \sum_i \sqrt{j_i(j_i+1)} \tag{33}$$

where γ is the Immirzi parameter [21]. We here observe that when the non-Abelian electric field is considered to be generated by the rotation of the non-Abelian magnetic field associated with the direction vector we can consider a Lorentz transformation such that from the frame of the background the magnetic flux will be transformed to the system moving with velocity v with respect to the background when we can write $E_a^i \sim \gamma v B_a^i$. Taking the factor $\gamma v = \alpha c$, c being the velocity of light we can replace E_a^i by αB_a^i. Taking α to be of the order of unity we have an equivalent relation for the unit of area A_0 in (30) as

$$A_0 = \alpha \mathscr{G} \hbar B(S), \quad S \in \Sigma \tag{34}$$

If N corresponds to the number of magnetic flux lines associated with the field $B(S)$ we note that the area unit is determined by the number of magnetic flux lines trapped within it. For a sequence of area bits $\Sigma^{(n)} = \Sigma_k^{(n)}, \quad k = 1 \cdots n,$ we can write

$$A(\Sigma) = \alpha \mathscr{G} \hbar \sum_{i=1}^{n} N_i \tag{35}$$

where N_i is the number of magnetic flux lines associated with the ith area bit. Thus we can write for the area in surface

$$A(\Sigma) = \alpha \mathscr{G} \hbar \sum_{i=1}^{n} N_i = \tilde{\alpha} \, \mathscr{G} \, \hbar \, n \tag{36}$$

n being the number of area bits and the unit of area A_0 is given by $\tilde{\alpha} \mathscr{G} \hbar$ in the theory. This helps us to write any area in a surface as

$$A = A_0 n \tag{37}$$

References

1. B.S. DeWitt, Quantum theory of gravity, Phy. Rev., 160 (1967), 1113.
2. S. Deser, N. Van Nieuwenhuizen and D. Boulware, General relativity and gravitation (Ed. G. Shaviv and N. Rosen), Wiley, New York, 1975.
3. A. Ashtekar, New variables for canonical gravity, Phys. Rev. D, 36 (1987), 1587.
4. R. Arnowitt, S. Deser and C. W. Misner, Gravitation: An introduction to current research (Ed. L. Witten), Wiley, New York, 1962.
5. J. F. Barbero, Real Ashtekar variables for Lorentzian signature space-times, arXiv: qc/9410014v1, 1994.
6. P. Bandyopadhyay and K. Hajra, Stochastic quantization of a Fermi field: fermions as solitons, J. Math. Phys., 28 (1987), 711.
7. E. Nelson, Dynamical theories of Brownian motion, Phys. Rev., 150 (1966), 1079.
8. K. Hajra and P. Bandyopadhyay, Equivalence of stachastic and klauder quantization and the concept of locality and non-locality in quantum mechanics, Phys. Lett. A, 7 (1991), 155.
9. P. Bandyopadhyay, The geometric phase and the spin-statistics relation, Proc. Roy. Soc. (London) A, 466 (2010), 2917.

10. R. Loll, New loop approach to Yang-Mills theory, Proceedings of the XIX International Colloquium on Group Theoretical Methods in Physics (Eds. A. Delolmo, M. Santader, J. Mateos-Guilart), vol. II (1992), 122.

11. T. Thiemann, Quantum gravity as the natural regulator of matter quantum field theories, Phys. Lett. B, 380 (1996), 257.

12. C. Rovelli and L. Smolin, Loop space representation of quantum general relativity, Nucl. Phys. B, 331(1990), 80.

13. Y.S. Wu and A. Zee, Geometric phases in physics, Nucl. Phys. B, 258 (1985), 157.

14. K. Sen and P. Bandyopadhyay, A geometrical formulation of Abelian gauge structure in non-Abelian gauge theory and disconnected gauge group, J. Math. Phys., 35 (1994), 2270.

15. D. Banerjee and P. Bandyopadhyay, Topological aspect of a fermion, chiral anomaly and Berry phase, J. Math. Phys., 33 (1992), 990.

16. L. Smolin and C. Soo, The chern-simons invariant as the natural time variable for classical and quantum cosmology, Nucl. Phys. B, 449 (1995), 289.

17. R.F. Bardeen, D.K. Sharp and J.A. Wheeler, From 3-geometry transition amplitudes to graviton states, Phys. Rev., 126 (1962), 1864.

18. K. Kuchar, A bubble-time canonical formalism for geometrodynamics, J. Math. Phys., 13 (1972), 768.

19. L. Mullick and P. Bandyopadhyay, Quantum geometry and entanglement entropy of a black hole, Gen. Rel. Grav., 44 (2012), 1199–1205.

20. C. Rovelli, Quantum Gravity, Cambridge University Press, 2004.

21. E. Verlinde, On the origin of gravity and the laws of motion, arXiv: 1001.0785, 2010.